Horse behaviour and welfare

EAAP – European Federation of Animal Science

The European Association for Animal Production wishes to express its appreciation to the *Ministero per le Politiche Agricole e Forestali* and the *Associazione Italiana Allevatori* for their valuable support of its activities

Horse behaviour and welfare

EAAP publication No. 122

Editors:

M. Hausberger, E. Søndergaard and W. Martin-Rossset

Wageningen Academic
Publishers

ISBN 978-90-8686-033-3

ISSN 0071-2477

First published, 2007

Wageningen Academic Publishers
The Netherlands, 2007

Table of contents

Foreword 9

Part A -Temperament / personality traits and genetics of behaviour in horses

Introduction of a behavioural test for Franches-Montagnes horses 13
D. Burger, I. Imboden, L. Jallon, J.-C. Ionita, V. Rapin, M. Doherr and P.-A. Poncet

Interplay between environmental and genetic factors in the behaviour of horses 23
M. Hausberger, M.A. Richard-Yris and A. Ricard

Part B - Early experience, human-foal relation and the impact of training

Early experience and behavioural development in foals 37
M. Hausberger, S. Henry and M.A. Richard-Yris

Effects of period, type and duration of handling on manageability, reactivity and learning ability of horses 47
L. Lansade, X. Boivin and M.F.Bouissou

Investigations on the activity of trotters and the relationship to growth 57
L. Voswinkel, K.-H. Tölle, C. Henze, K. Blobel and J. Krieter

Part C - Importance of social environment on the behaviour of young and adult horses

The effects on behaviour of the presence or absence of a familiar horse during feeding 65
S.E. Redgate and H.P.B. Davidson

The effects of social environment on the behaviour of young horses 71
E. Søndergaard and J.W. Christensen

Individual paddocks versus social enclosure for horses 79
G.H.M. Jørgensen and K.E. Bøe

Part D - Equine behaviour welfare

The management of horse behaviour in captivity and its welfare implications 87
D.S. Mills

Impact of Nutrition and Feeding practices on equines, their behaviour and welfare 99
P.A. Harris

Three-dimensional design of a horse stud like better toll for technical choices of housing and welfare 115
N. Miraglia and A. Simoni

Behavioural repertoire: its expression according to environmental conditions 123
H. Benhajali, M. Hausberger and M.A. Richard-Yris

Building and process technology requirements in horse husbandry systems: current investigations and developments furthering animal welfare and environmental protection 139
F.-J. Bockisch, P. Kreimeier, G. Hoffmann, T. Hohmann, W. Bohnet and U. Brehme

Foreword

Extensive attention has been paid to study Genetics, Physiology, Nutrition, Pathology of Equines as the practical issue of the research can be directly and quantitatively evaluated: fertility, efficiency of utilisation of nutrients, performance, health, cost ... But it has been clearly highlighted in other farm animals that a better knowledge of the behaviour and of the main components of the welfare could contribute to the improvement of the productivity of animals and the quality of the animal products.

Equines are mainly sport and recreational animals as they are used for racing, riding, driving or drafting, ... Equines are bred and used in a very large variety of situations since the early time. The duration of utilisation of equines averages approximately 10 years. As a result a lot of husbandry, education, training techniques and devices have been designed. Particular relationship is set up between equines and owners. And ultimately the equines can be companion animals of the owners, which have, or not enough background to manage them properly. Since long time Equines are taking part of the socio-cultural life as well in rural as in urban environment. Owing to recent evolution in the industrialised countries, breeding and utilisation of the Equines in the various situations are influenced by the public opinion. As a result additional attention has to be paid to the welfare of the Equines as well for recreational use as for competing.

The Horse commission of the European Association for Animal Production (EAAP) uses to perform, in the scope of annual meetings of EAAP, scientific sessions devoted to disciplinary or multidisciplinary approach to answer to questions arising from Equine industry and Society. Working groups have been set up too under the umbrella of the Horse commission to examine thoroughly the scientific thought and to contribute to the settlement of a European network for research and harmonisation of methodology and tools to be implemented in Equine industry at EU level, and to promote consistent dissemination of knowledge to end users. In this context a session dedicated to the study of behaviour and welfare has been performed at the 54th annual meeting held in Rome in 2003. A working group has been set up in the field. This group is led by Eva Sondergaard from Denmark, Martine Hausberger from France and Mark Kennedy from United Kingdom. A first summer school of Ethology of Horses has been organised in France in 2005 by the University of Rennes in Brittany under the umbrella of the French Ministry of Education and Research and EAAP; the event has been chaired by Martine Hausberger. And in 2006 at the 57th annual meeting held in Antalya in Turkey a session was dedicated to the effect of management and housing on horse welfare.

This special issue aims at publishing the main papers in the field of temperament, early experience, social management and equine behaviour welfare. Precise studies performed in these fields are described here, that are placed in the more general context of the international scientific advances.

Two papers deal with temperament and the issues of selection of behaviour. One attempt of testing behavioural tests in a selection program is described by Burger *et al.* and reveals both the advantages and difficulties in such attempts. It appears clearly that the use of usual working tasks raises the issue of learned (trained) versus more genetic trends. Hausberger *et al.* describe a large scale study on temperament that reveals influences of a variety of factors, amongst which the breed, sire and type of work. A first calculation of heritabilities suggests an interest in developing further such approaches.

The two papers that deal with early experience and handling of foals (Hausberger *et al.* and Lansade *et al.*) show that the issue is complex, as the final results depend on the type of handling, that can induce positive as well as negative effects on the human/foal relation, the time and duration of handling. An emphasis is given on the use of a social model (the dam) and the weaning period as two aspects that can facilitate the establishment of a positive human/foal relation. Early experience in terms of housing and exercise can also have an impact on growth and training as shown by Voswinkel *et al.*

Social environment again can have good and bad aspects and an appropriate management is required. Thus, the presence of another horse while feeding may be either beneficial or potentially stressful while an appropriate social environment is required in order for the young horses to develop and practice their social skills, as demonstrated by Redgate *et al.*, Søndergaard *et al.* and Jørgensen *et al.*

The importance of management for the horse's welfare is demonstrated by Mills' paper that gives a general overview of our current knowledge in the field. Amongst the important aspects are que questions of nutrition (Harris) and housing (Miraglia *et al.* and Bockisch *et al.*) but also the possible behavioural indicators of stress and/or well being (Benhadjali *et al.*).

All aspects of this special issue therefore are complementary and as a whole contribute to the human/horse relation and to both human and horse welfare.

New issues are raised that suggest other ways in the human/horse interaction: an awareness of individuality in behaviour, of the influence of the social environment, on when and how to establish properly the relatio. A promising program for fundamental and applied research and daily practices.

Martine Hausberger, Eva Sondergaard and W Martin-Rosset

Part A

Temperament / personality traits and genetics of behaviour in horses

Introduction of a behavioural test for Franches-Montagnes horses

D. Burger[1], I. Imboden[1], L. Jallon[2], J.-C. Ionita[1], V. Rapin[1], M. Doherr[3] and P.-A. Poncet[1]*
[1]Swiss National Studfarm, 1580 Avenches, Switzerland, dominique.burger@mbox.haras.admin.ch
[2]Swiss Franches-Montagnes Breeding Association, 1580 Avenches, Switzerland
[3]Department for Clinical Veterinary Medicine, University of Bern, 3001 Bern, Switzerland

Abstract

A horse's character and behaviour are of the greatest importance for its day to day handling and use in leisure and in sporting activities. To better provide for the needs of the clients of the Franches-Montagnes breed - mostly amateur pleasure riders and drivers - routine official behaviour testing for these horses was introduced in 2001. The tests are carried out in conjunction with the breed association's field tests for three-year-old Franches-Montagnes horses and the annual 40-day station test for young approved stallions. The testing procedure for the field tests involves the application of one randomly selected test out of four possible tests on the standing horse, one out of three tests on the ridden horse, and one out of four tests on the driven horse, while at the station tests, all tests are carried out. The horses' reactions are assessed and graded by previously trained experts. Some of the tests resemble already existing, well-documented methods that have been adapted to field conditions, some have been specially developed for the purposes of the behavioural test. Test procedures, their sensitivity and repeatability as well as the results obtained from tests on 2285 three-year-old and 35 approved Franches-Montagnes stallions, 22 Trotters and 24 Warmblood horses are described and analysed. The results show some significant differences and low correlations between both the individual test results and between the test group results. They have provided a large data-base for the analysis of the temperament of the Franches-Montagnes horse and for a basis for comparison between different breeds. First experiences gathered on the introduction of a behavioural test on the occasion of a breeding selection program and the effects and consequences for the breeders and clients are illustrated and discussed from both a scientific and a political point of view.

Keywords: behavioural tests, horse, emotivity, selection, temperament

Introduction

Along with conformation, gaits and performance of a horse, its character and behaviour are of the greatest importance, not only in companion animals and for leisure activities, but also in sport horses. There is an alarming increase in the number of accidents with horses, due to the lack of experience and skills of the amateur rider and characteristic behaviour patterns of the horse (Hobbs *et al.*, 1994; Keeling *et al.*, 1999). In this context there is ever increasing interest in the study of the temperament of horses as well as in the possibility of a character based selection process.

Behavioural and character traits of a horse are, however, very difficult to define and measure objectively. The terms 'flighty' and 'sharp' are the only commonly understood descriptions of a horse's behaviour in the equine industry (Mills, 1998). Nevertheless, Le Scolan *et al.* (1997) found that experimental tests may be good predictors of temperament in untrained horses; Visser *et al.* (2003) were able to demonstrate that it is possible for a large panel of assessors to agree upon a horse's temperament and that objective measurements from behavioural tests correlate significantly with temperamental traits evaluated by a panel of assessors. Up until now, most of the research carried out in the field of behavioural tests has focused on the so called 'emotivity' of the horse: traits such as fearfulness and gregariousness (Vierin *et al.*, 1998; Wolff *et al.*, 1997) which are of great importance in the day to day handling of horses. The influence of genetic and environmental factors on these character traits has been studied by Hausberger *et al.* (1996; 1998) and Vierin *et al.* (1998). They found that both a horse's breed and its sire may influence the behavioural reactions to the tests and play an important role in its reactions to new objects. Environmental factors, in particular the type of work that has been done with the horse, modify the effects of this genetic basis.

To better provide for the needs of the clients of the Franches-Montagnes breed - mostly amateur pleasure riders and drivers - routine official behaviour testing for these horses was introduced in 2001 by the Swiss Franches-Montagnes Breeding Association. The present study describes both the scientific and political aspects of the introduction of an emotivity orientated behavioural test into a breed selection program. It also briefly compares the temperament of the Franches-Montagnes breed with that of two other horse breeds, the Trotter and the Warmblood.

Materials and methods

Animals and procedures

Eleven tests were developed: four tests on the standing horse, three on the ridden horse and four on the driven horse. Most of the tests resemble already existing, well-documented and well-established methods to measure reactions to new objects, but have been adapted to the conditions in the field (tests should be: short in duration, requiring little in the way of infrastructure and expertise, easily integrated into an existing field test).

The tests were carried out in conjunction with the annual 40-day station tests for the young breed-association-approved stallions (a); in conjunction with the breed association's field tests for three-year-old Franches-Montagnes horses (b); and separately as breed tests (c).

a: station tests (ST)

35 breed society approved three-year-old Franches-Montagnes stallions were tested at the 40-day station test for the approved stallions in 2002 and 2003 in the Swiss National Studfarm in Avenches. All eleven tests were performed twice in an interval of two weeks.

b: field tests (FT)

2285 three-year-old Franches-Montagnes mares and geldings were tested at the breed association's field tests in 2001, 2002 and 2003 in a total of 38 (5 - 72 horses per event), 34 (4 - 99 horses per event) and 38 locations (7 - 69 horses per event) respectively. The testing procedure involved the application of one randomly selected test out of each of the categories (standing, ridden and driven). The breed association decided in advance which of the tests were applied in each different field test location.

c: breed tests (BT)

Twelve three-year-old Franches-Montagnes mares and geldings, 22 two- and three-year-old Trotters (all sexes) and 24 three-year-old Warmblood horses (all sexes) were tested and compared in 2003. The tests were carried out at the Swiss National Studfarm in Avenches, the National Horse Center in Berne and at a stud in Payerne (Olivier Pradervand/ Gerhard Etter). All four tests on the standing horse were performed.

Tests

Tests on the standing horse

The horse is held loosely in quiet surroundings, wearing either a headcollar or a bridle. The person holding the horse must not give any verbal aids that might influence the horse.

Ear test: The expert approaches the horse from the left. He halts at shoulder level half a meter away from the horse and gently strokes the horse's neck with his right hand. He then gently takes hold of the horse's left ear with his right hand and holds it for 2 seconds. If this manoeuvre is successful, the procedure is repeated with the right ear, which is also held for 2 seconds. If the horse resists the manipulation one more attempt is allowed.

Hoof test: The handler (horse-owner) approaches the horse from the left side, runs his right hand down the horses neck and gently picks up the left fore. He then runs his hand over the horse's croup and picks up the left hind. The procedure is repeated from the other side with the right fore and hind. While holding the hoof he uses a hammer to gently tap the inner and outer quarters of the horse's shoe or hoof wall (if it is unshod).

Umbrella test: The expert approaches the horse from the left side and halts 5 meters from the horse. He opens an umbrella and closes it again immediately. The test is repeated at a distance of 3 meters.

Spray test: The expert approaches the horse from the left side and halts 3 meters from the horse. He sprays water from a commercially available plant spray bottle in the horse's direction for 2 seconds, then approaches to a distance of half a meter and sprays directly at the horse's neck for 2 seconds.

Tests on the ridden horse

Tests on the ridden horse are carried out in an indoor or outdoor arena. The horse must be ridden on a loose rein and has 90 seconds of time to complete the task. The approach begins at a distance of 10 meters, for the first 5 meters the rider may use aids to drive the horse towards the obstacle. After this point he has 30 seconds in which he may use neither verbal nor leg aids, after 30 seconds he may use only his voice, and after 60 seconds he may use his legs to drive the horse forward.

Umbrella test: Two colourful umbrellas are firmly fixed to the ground one-meter apart. The rider has to steer the horse through the gap.

Ball test: A large rubber ball (1 meter in diameter) is placed on the ground. The rider attempts to ride the horse up to the ball and to touch it in some way.

Ground test: A strongly contrasting area of ground is created, using coloured sand or chalk. The rider attempts to ride the horse over this obstacle.

Tests on the driven horse

Tests on the driven horse are carried out in an outdoor driving square. The driver must attempt to steer the horse through the obstacle in the 90 seconds of time available for this task. The rules for the approach are the same as for the ridden tests, instead of leg aids, the whip may be used.

Curtain test: A curtain of plastic strips 4 meters high and 4 meters wide is erected in a suitable location. The plastic strips are attached at intervals of 20 cm and are long enough to touch the ground. The driver attempts to drive the horse through the curtain.

Umbrella test: As for ridden test, but with a gap of 2.5 m between the umbrellas.

Ground test: As for ridden test.

Corridor test: A fenced corridor 8 meters long and 2.5 meters wide is erected on a driving square. Ropes are fastened along both sides of the corridor to which fluttering strips of plastic are attached at intervals of 20 cm. One of the ropes can be manipulated, using an attached piece of string, by a helper standing 4 meters away from the corridor. The driver must attempt to steer the horse through this corridor.

Where testing took place in an outdoor-school, it was not possible to carry out the tests which involved either an umbrella or a ball in very windy conditions.

Scoring

The horses' reactions were assessed with a scaling protocol and graded using a scale of 1 (unsatisfactory/ strong reaction) to 9 (excellent/ no reaction) The horses each got an overall behaviour score, calculated from their average marks. The ST were always judged by the same two experts, the FT by one of a total of 33 previously instructed experts and the BT always by the same expert.

Statistical analysis

The results were evaluated and compared by statistical means. Non-parametric statistics were used as the data did not follow a normal distribution. Significance was determined at the $P < 0.05$ level. The tests were analysed according to the following criteria: procedure, personnel and infrastructure requirements, duration and risk for the persons involved, scoring system, sensitivity, specific character traits tested, repeatability and comparison between breeds.

Results

A total of 2356 horses have been tested at the station tests, the field tests and the breed tests. Scores ranged from 1 to 9, the average score obtained in a total of 6943 tests was 8.19 (sd ± 1.57). Only 495 horses (7.13%) achieved a mark of ≤ 5 (Figure 1).

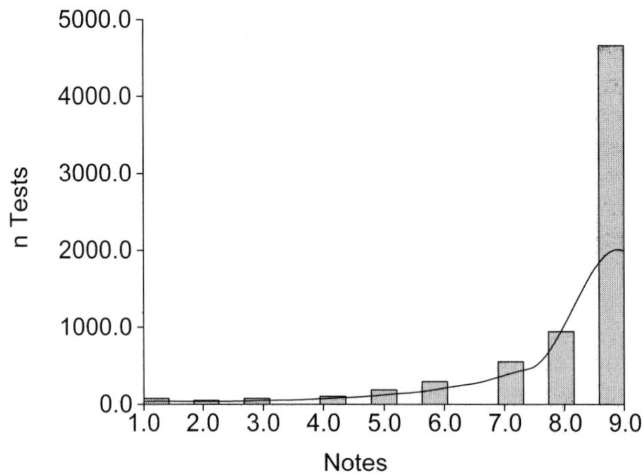

Figure 1. Histogram of all marks obtained in a total of 6943 behavioural tests.

Comparison between the test group results (FT, ST & BT)

Significant differences exist between the test groups: while stallions in the station tests 2002 and 2003 obtained an average score of 7.14 (7.41 in the tests on the standing horse), horses in the field tests 2001 - 2003 achieved 8.25 (8.36 in the tests on the standing horse) and the Franches-Montagnes horses in the breed tests 2003 on the standing horse 8.04 (Figures 2 and 3). Comparing the results of the standing horse tests of Franches-Montagnes horses in the ST, the FT and the BT using the Kruskal-Wallis multiple-comparison z-value test, showed significant differences of the scores in the ear and spray test (comparing ST with FT) as well as in the hoof and umbrella test (ST - FT and ST - BT). In the driving and riding tests, Mann Whitney or Wilcoxon rank-sum tests revealed very significant differences between the test groups ($P< 0.005$) with the exception of the corridor test for driven horses ($P< 0.05$) and the ball and ground test (no difference) under the rider.

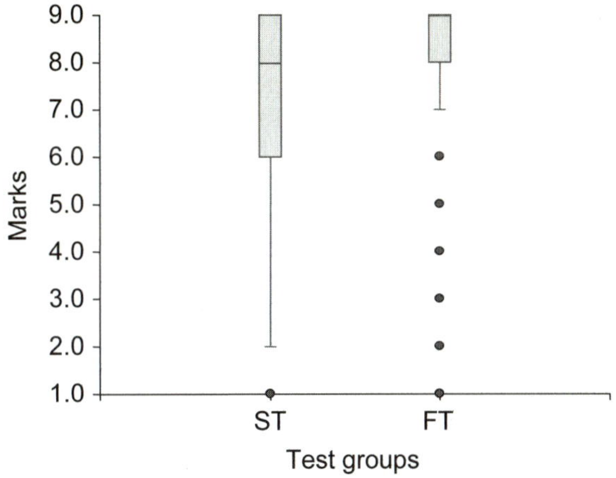

Figure 2. Marks (box plots) of all tests on the standing, ridden and driven horse in the station tests (ST) and field tests (FT).

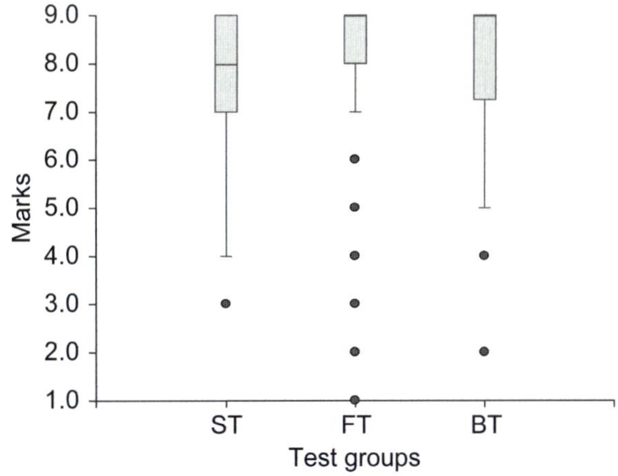

Figure 3. Marks (box plots) of all tests on the standing Franches-Montagnes horses in the station tests (ST), field tests (FT) and breed tests (BT).

Comparison between the individual test results in the station tests

The performing of all available tests at the 40-day station tests allowed evaluation and comparison of the individual results. This gave some indication about the severity of the different tests (a test with a high median score might have been easier to perform than one where the average score was lower) and provided an indication about the degree in which they can be directly compared with each other. Figure 4 shows the mean values obtained in all tests. The Kruskal-Wallis multiple comparison z-value test showed that tests have varying median mark levels. This is especially pronounced in the driving tests, where 2 of the 4 tests resulted in significantly higher median values (Bonferroni test z-values of > 3.3172).

Spearman rank analysis allowed us to evaluate the extent of correlation between the results achieved by each individual in each of the performed tests. A high correlation would indicate that a horse's results were fairly constant, while a low correlation would result from horses obtaining a variety of scores for the different tests; either due to inconsistency in the horse's behaviour, or due to the tests evaluating different traits of the horse's character. The majority of the calculated correlation coefficients between the test results obtained in the 40-day station tests (first tests per stallion) were statistically non significant ($P > 0.05$) low to very low (-0.350 to 0.350). Exceptions of this are found comparing the ground tests on the ridden horse with the curtain tests on the driven horse (0.702; $P < 0.001$), the spray tests on the standing horse with the ground tests on the driven horse (0.477; $P < 0.005$), the ball tests on the ridden horse with the umbrella tests on the driven horse (0.475; $P < 0.05$) and the umbrella tests on the ridden horse with the umbrella tests on the driven horse.

Repeatability

The repeatability of the tests was statistically analysed by a Spearman rank correlation test using the first and the second test series carried out at the 40-day station test for the young approved stallions. Values varied between 0.249 (low correlation) and 0.720 (high correlation) with a mean of 0.496 (Table 1). With the exception of the standing ear test and the ridden ground test all tests resulted in significantly ($P < 0.001$) higher scores in the second test series (mean value of 7.48 vs. 7.14). In the driven corridor test, all horses obtained a maximal score of 9 in the second series of tests; therefore statistical analysis of this test was not possible.

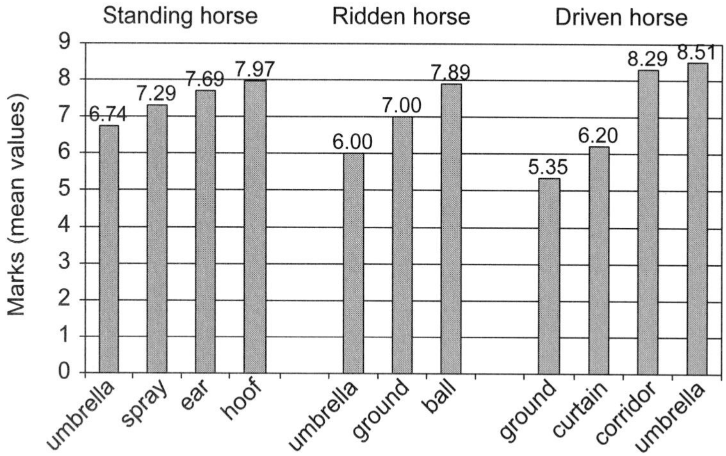

Figure 4. 40-day station tests for the young approved stallions 2002 and 2003 (first tests per stallion): mean values of all tests on the standing horse, the ridden horse and the driven horse.

Table 1. Repeatability of behavioural tests in 35 stallions.

Test	Repeatability	Mean
Driven curtain test	0.450	
Driven umbrella test	0.501	
Driven ground test	0.359	
Driven corridor test	-	
Driven tests: mean		0.480
Standing ear test	0.701	
Standing umbrella test	0.322	
Standing hoof test	0.494	
Standing spray test	0.262	
Standing tests: mean		0.415
Ridden umbrella test	0.251	
Ridden ball test	0.249	
Ridden ground test	0.720	
Ridden tests: mean		0.554
All tests: mean	0.496	

Comparison between the breeds

The results of the four tests on the standing horse carried out in the breed tests on 12 three-year-old Franches-Montagnes horses, 22 two- and three-year-old Trotters and 24 three-year-old Warmblood horses were compared using a Kruskal-Wallis multiple-comparison z-value test. Significant differences were found in the ear test (Trotter - Warmblood horses) and in the hoof test (Franches-Montagnes horses - Warmblood horses). Comparing all tests performed, the Warmblood horses obtained significantly lower marks than the other two breeds (Figure 5).

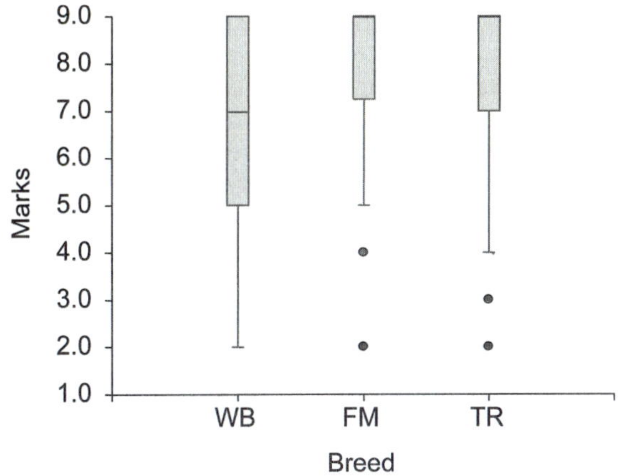

Figure 5. Marks (box plots) in all tests on the standing horse of 24 three-year-old Warmblood horses (WB), 12 three-year-old horses (FM) and 22 two- and three-year-old Trotters (TR).

Discussion

In the Swiss horse industry, the Franches-Montagnes horses are generally considered to be very even tempered; a fact often used when promoting the breed. The breed association's aim, when introducing these tests, was to highlight this quality while at the same time eliminating horses with unacceptable behavioural traits from the breeding program.

Scientific evaluation

For reasons of practicality, the behavioural tests had to be included in the already existing approbation programs such as the station test for young stallions and the field test program for 3-year-old horses. They had to have minimal infrastructure and personnel requirements, be short in duration and be straight forward enough to be executed and assessed by 33 different experts. The behavioural tests took approximately 3 minutes each, and in the case of the field tests (which provided by far the largest number of horses) were carried out in conjunction with a conformation, ridden and driven test.

In general the scores achieved were very high, which would suggest that the tests are not severe enough. However, as the breed tests demonstrated, there is a statistically significant difference between the breeds: Warmblood horses appear to have a demonstrably higher emotivity than Franches-Montagnes horses and trotters. Various authors have already described differences in behavioural reactions to experimental situations in animals issuing from different sires or belonging to different breeds (Hausberger and Ricard, 2002).

The comparison between the test groups (field tests, station tests and Franches-Montagnes horses in breed tests) showed significant differences between the mean scores achieved by the different groups. Various factors could be responsible for this phenomenon:

- differences between the sexes - station tests are restricted to stallions, whereas field and breed involved only mares and geldings;
- differences in quality of judging;
- differences in the education and training level of the horses;
- environmental influences - station and breed tests were carried out in surroundings familiar to the horse, while the field tests took place in unfamiliar surroundings.

It is unlikely that the sex of the horses had a large influence - a study by Hausberger *et al.* (1998) showed that emotivity specific tests are not significantly influenced by this factor. It seems more likely that the difference occurred in the execution and interpretation of the tests. In the field tests, carried out in the presence of the horse's owner, the testing experts are the conformation, riding and driving judges. These were individuals working on a voluntary basis for the breed association and were instructed in the specific skills required for the judging of the behavioural tests, including a clear scaling protocol, during the annual courses organised by the breed association. These experts tended to award higher marks than those in the station test. The station and breed tests, which were carried out without the presence of the horse owner and always by the same experts, probably allowed a more critical evaluation.

The comparison between the individual tests of the horses in the station tests showed significant differences between the average test scores. Correlations between an individual's test results were in general low. The differences between the test scores could indicate that not all tests are of equal severity. However, when combined with the low correlation seen when an individual's test scores are statistically compared with each other, it appears more probable that not all tests evaluate the same character trait or that they have a low specificity for any particular character trait. These aspects have also been described by Le Scolan *et al.* (1997) as well as Hausberger and Ricard (2002) concluding that emotivity, tested in different situations, presents a multifactorial challenge. The hoof test, for instance, tests a horse's amenability, while the umbrella test is an indicator of its flightiness. Specific evaluation of the tests in this context remains necessary.

Repeatability of the tests was very variable and with the exception of the standing ear test and the ridden ground test, not high. This could indicate that some of the tests used are not an appropriate indicator of a horses' true character, but rather a reflection of its level of training and obedience. The driven corridor test, for instance, is an obstacle routinely used in the education and training of the young stallions during the station test. The risk of habituation while repeating tests is well known (Jones, 1989). Correlation of successive measurements of emotivity - assessing the stability of a horses temperament - is difficult to evaluate in adult horses (Hausberger and Ricard, 2002; Visser *et al.* 2001). Another important factor affecting the results was the level of training and quality of handling the 3 years olds had received from their breeders.

Political viewpoint and experiences

The first experiences gathered on the introduction of a behavioural test on the occasion of a breeding selection program and the effects and consequences for the breeders and customers are positive. The test's introduction was well received by both breeders and leisure riders. Initially breeders were concerned that the tests would add to the selection pressure of the young horses; however the fact that the scores were generally high and that many potential clients showed great interest in a horse's behavioural test score helped to gain breeders' approval.

Many breeders have started training their young horses specifically for the behaviour test; for example stallions may have a ball in their box, umbrellas are placed in the stables and most breeders condition their horses to the use of a spray. Although difficult to evaluate from a scientific point of view, empirically, it can already be clearly seen, that education of young Franches-Montagnes horses has evolved in a positive direction.

The most criticism came from the part of Switzerland where Franches-Montagnes horses are traditionally kept very extensively, but after one year of acclimatisation the behaviour test is now also accepted in this region. Initial concerns were voiced about the safety of a couple of tests, however, it must be noted that in the course of all the tests performed in this study, there was no accident or injury of either horses or people.

Problematic remains the fact, that only emotivity is tested. Not desired character traits as aggressiveness are not involved in the judgement and as a consequence an inappropriate assessment of a horse's general temperament may result.

Conclusion

The first introduction of official behaviour testing of all young horses in a breeding program has given predominantly satisfying results. Though scientifically difficult to evaluate and to some extent lacking in precision, the emotivity of Franches-Montagnes horses has been successfully assessed and documented. As a result of these tests, breeders are starting to put more weight on the handling of their young animals. This leads to a better satisfaction of potential customers and the public in general.

Behavioural testing will also be an important instrument for stallion selection and breeding strategies. It is assumed, that in the future, it will be possible to statistically evaluate the previously approved stallions according to the behaviour test results of their progeny. The behavioural tests carried out in the station test are already serving as a parameter in the selection of new approved stallions.

To optimise the behavioural tests' fairness and ease of interpretation, the choice of test at the field tests must be done with more respect for the character trait being tested, the trainability and the severity of each test. Moreover the behavioural tests should not only be emotivity-orientated. In the next years, some of the tests will be replaced by more differentiated tests, or the grading systems will be slightly changed. Replacing the grading system with a simple pass or fail or a 'linear description' is also being considered. Due to the restricted time at disposal during the field test, more detailed and therefore more time consuming testing would not be possible in the field. The instruction of the experts and the quality of judging will be reinforced to obtain better standardisation and more objective testing.

In the future, we also intend to expand our work with temperament evaluation of Franches-Montagnes horses by introducing a detailed behavioural test for horses belonging to horse-owners interested in partaking in such studies. This would take place over two days and would allow a much more accurate identification and assessment of a horse's character traits as well as its capacity for learning and memorisation under more standardised conditions.

Acknowledgements

We would like to thank all horse owners involved in the testing, especially the National Horse Center in Berne and Gerhard Etter, Müntschemier, for putting their horses at our disposal.

References

Hausberger, M., N. Le Scolan and C. Bruderer, 1998. Le temperament du cheval: Facteurs en jeu et implications practiques. Recherche Equine 24, 159-170.

Hausberger, M., N. Le Scolan, C. Muller, E. Gautier and A. Wolff, 1996. Caractéristiques individuelles dans le comportement du cheval: prédictibilité, facteurs endogènes et environnementaux. In: Institut du Cheval, DEFI, 22ème Journée d'Etude, 113-123.

Hausberger, M. and A. Ricard, 2002. Génétique et comportement chez le cheval. INRA Prod. Anim. 15 (5), 383-389.

Hobbs, G.D., D.M. Yealy and J. Rivas, 1994. Equestrian injuries: a five-year review. J. emerg. med. 12, 143-145.

Jones, R.B., 1989. Avian open-field research and related effects of environmental novelty: an annoted bibliography, 1960-1988. Psychol. Rec. 39, 397-420.

Keeling, L.J., A. Blomberg and J. Ladewig, 1999. Horse-riding accidents: when the human-animal relationship goes wrong! In: 33rd International Congress of the International Society for Applied Ethology.

Le Scolan, N., M. Hausberger and A. Wolff, 1997. Stability over situations in temperamental traits of horses as revealed by experimental and scoring approaches. Appl. anim. behav. Sci. 41, 257-266.

Mills, D.S., 1998. Personality and individual differences in the horse, their significance, use and measurement. Equine vet. J., Suppl. 27, 10-13.

Vierin, M., M.F. Bouissou, M. Vandenheede, C. Trillaud-Geyl and G. Arnaud, 1998. Développement d'une méthodologie destinée à mesurer les réactions de peur chez le cheval. Application à l'étude de l'influence de différents facteurs. In: Institut du Cheval, DEFI, 24ème Journée d'Etude, 1-15.

Visser, E.K., C.G. Van Reenen, H. Hopster, M.B.H. Schilder, J.H. Knaap, A. Barneveld and H.J. Blokhuis, 2001. Quantifying aspects of young horses'temperament: consistency of behavioural variables. Appl. anim. behav. Sci. 74, 241-258.

Visser, E.K., C.G. Van Reenen, M. Rundgren, M. Zetterqvist, K. Morgan and H.J. Blokhuis, 2003. Responses of horses in behavioural tests correlate with temperament assessed by riders. Equine vet. J. 35, 176-183.

Wolff, A., M. Hausberger and N. Le Scolan, 1997. Experimental tests to assess emotionality in horses. Appl. anim. behav. Sci. 40, 209-221.

Interplay between environmental and genetic factors in the behaviour of horses

M. Hausberger[1], M.A. Richard-Yris[1] and A. Ricard[2]
[1]Université Rennes 1, UMR 6552 Ethologie - Evolution - Ecologie, Campus de Beaulieu, 35042 Rennes Cedex, France
[2]SAGA, Toulouse, France

Abstract

Behavioural genetics of horses are in their infancy. The importance of taking behaviour into account during selection appeared only recently. In addition, this is a difficult question to address for this species as the living conditions of these animals are very varied. We have no data on genetic determination, only indirect indications based on comparisons among breeds or lines. Greater similitudes among animals with the same father and/or of the breed were observed in foals, as well as in adults, in freely expressed activities (play, distance from mothers) and in reactions to experimental situations. Stereotyped activities appear to be related, in addition to environmental conditions, to family and to breed. The relative weights of environmental and genetic factors on the determination of behavioural characteristics appear to differ in relation to the character considered. The differences observed between families or breeds could also be linked to epigenetic factors (maternal effect for example). Taking behaviour into account when selecting horses appears advisable and possible, on condition that the character considered is well defined, its measurement is standardised and the impact of epigenetic factors is evaluated.

Keywords: horse, temperament, behaviour genetics

Introduction

As Bateson (2000) mentioned, 'the development of behaviour clearly depends both on inherited factors (primarily genes) and non-inherited factors (primarily environmental influences)'. One can therefore not ask whether a behaviour is genetic or is learned but rather what kind of influences a variety of factors, either genetic or epigenetic, can have on individual behaviours.

Evidence for genetic bases to behaviour has been found in a variety of species like quails (Faure and Mills, 1998), foxes (Belyaev and Trut, 1975), cattle (Fordyce *et al.*, 1982, Le Neindre *et al.*, 1995) or dogs (Goddard and Beilharz, 1982) amongst domestic species. Mc Cune (1995) looked at environmental modulation of genetic influences in kittens: the friendliness to humans in this species is related to the father but handling by humans can modulate this influence.

The situation in horses is much less well known. Genetics of horse behaviour are only just beginning to be studied (Houpt and Kusunose 2001), because, among other reasons no doubt, of the difficulties to find a sufficient number of related animals and of the diversity of environmental situations.

Selections based on performance or morphology have yielded results (Langlois, 1984; Ricard, 1997) and for some time, many reports have recommended that behaviour of horses should be included in the selection criteria as some behavioural traits are particularly important for the horse's use. In German selection appreciations are given of 'blood', character, aptitude to be ridden and 'heart' of young horses (Bruns, 1982). The general problem in recent years has been that these behavioural evaluations were rather subjective (Langlois, 1984).

Most of the data concerning possible genetic bases or even mere individual variation of horse behaviour are recent. Individual variations can be evidenced by observing animals in their usual habitat. However, generally, the more fundamental tendencies related to genotype are revealed in unfamiliar situations (Gerlai and Csañyi, 1990). Selection possibilities are based on the postulate that behavioural tendencies are maintained over time and situations, and are, when possible, predictable at a young age (Hall, 1941; Bates, 1989). Currently, we can only review the existing data in horses that dissimilar methods or samples have yielded, but that inform about possible lines of research.

Data on genetic bases are mainly relying upon comparisons of breeds or bloodlines giving undirect evidence for a possible genetic transmission.

Evidence of genetic bases to horses' behaviour

Genetic bases of spontaneous activities

Observations of spontaneous activities under familiar conditions reveal that horses, as other species, present clear inter-individual differences.

In addition to observations of groups showing the existence of different ranking status among individuals (Tyler, 1972), other observations reveal inter-individual variations for other traits as different as food choice (Marinier and Alexander 1992), maternal behaviour and mare-foal distance. Differences in activity profiles and time-budgets have been observed among mares and foals (Tyler, 1972; Crowell-Davies, 1986; Wolff and Hausberger, 1994). These variations could be the result of different mechanisms as learning mother's social rank (Houpt and Wolski, 1980; Houpt *et al.*, 1982), variations of play among foals in relation to sex (Waring, 1983), etc. Possible genetic bases for these differences have been little investigated.

Behaviour of foals

Wolff and Hausberger (1994), in a preliminary study on 13 3-month old French Saddlebred foals observed in paddocks with their dams (under conditions limiting the weight of environmental factors: same housing, food, handling since birth), revealed differences between families for durations of rest periods, mare-foal distances and time spent playing. A similar study, made three years later at the same breeding farm, confirmed these observations. In addition, the same tendencies were observed for the same paternal origins in two generations of horses, at a three-year interval and on the basis of 32 foals (Hausberger *et al.*, 1996; Hausberger and Richard-Yris, 2005).

Abnormal behaviour

In a report based on a large scale survey, Juarbe-Diaz *et al.* (1998) indicated differences between two breeds (Purebred Arabs: 5% and Paint horses: 2% in the tendency for mares to reject their foals but also differences between families as two stallions were overrepresented in the origins of Arabian rejecting mares.

Hosoda (1950) was the first to suggest a genetic predisposition for wind-sucking. He showed that wind-sucking frequencies varied among families from 1 to 7%. Later, families of Italian Thoroughbred horses presenting a high level of wind-sucking were identified (Vecchioti and Galanti, 1986). Luescher *et al.* (1998) estimated the probabilities of developing any type of stereotypic activity at 0.60 when one of the two parents presents a stereotypy and 0.89 when both parents present a stereotypy.

Several studies indicate that Thoroughbreds are more likely to present stereotypic activities than Trotters or Saddlebreds, particularly concerning wind-sucking and weaving, which is also more frequent in purebred Arabs (Luescher *et al.*, 1998; Redbo *et al.*, 1998). As these results were drawn from surveys made at different farms, it is difficult to dissociate the effects of genetic factors from those of the environment. McGreevy *et al.* (1995) showed that Thoroughbreds often have less social contacts with other horses, spend less time out of their boxes and receive more concentrated food. Hausberger *et al.* (1996) reported that, at the Saumur French national riding school under homogeneous living conditions, Anglo-Arab horses (AA) presented more stereotypies and spent less time sleeping than French Saddlebreds when they were observed in their boxes. This was particularly true when both parents were Anglo-Arabs.

Finally, some breeds or families appear to be predisposed to self-mutilation by biting (Dodman *et al.*, 1994).

Temperament / personality traits

The need for experimental tests that evaluate underlying individual behavioural tendencies appeared at an early date (e.g. Hall, 1941). Several tests or measures have been developed to evaluate inter-individual variations in horses. Most of the studies have been focused on aspects of emotional reactivity, but strong inter-individual differences exist in aggression towards conspecifics or in social status determining the place of each individual within the social group (Waring, 2003). No appreciable differences among breeds concerning social behaviour or time budgets appear to have been reported.

Experimental tests

Two aspects (learning and emotionality) have been more particularly studied, as these traits are especially important in the horses' daily use.

Learning abilities: In one of the first early studies, horses had to discriminate four doors covered with different visual patterns to find food and, in this task, Quarter Horses appeared to be more successful than Thoroughbreds suggesting breed differences in learning abilities (Mader and Price, 1980). One can not exclude however that Thoroughbreds may have been more afraid of the testing environment.

Several studies based on maze tests revealed inter-individual differences in performance (Heird *et al.*, 1986). Budzynski *et al.* (1992) observed differences between breeds after studying over 400 horses belonging to three Polish breeds. Wolff and Hausberger (1996) presented a 'detour' task (horses had to find a bucket with food, placed in the center of a circle, the entrance of which was opposite the initial position of the horse) to 2-3 year old French Saddlebreds, all reared under the same conditions, and these authors observed the existence of inter-individual variations, some of which related to paternal origin.

In the same study, Wolff and Hausberger (1996) taught these young horses to open a wooden chest containing food. This test again evidenced paternal effects: the siblings of one stallion tended to learn faster than unrelated foals of other sires. The same test, carried out at the Saumur ENE revealed differences between French Saddlebred and Anglo-Arab geldings, the latter showing greater difficulty learning the task (Hausberger *et al.*, 1996). Finally, differences between breeds were observed in a large scale study including over 700 horses: Haflinger or Appaloosa horses learnt this task more quickly than Purebred Arab, Angloarabs or French Saddlebreds horses (Hausberger *et al.*, 1998, 2004).

Emotionality: Few experimental studies have explored breed or sire differences in emotional reactions to challenging situations. Wolff and Hausberger (1992) and Wolff *et al.* (1997) using three different experimental tests observed more similar reactions between half siblings than unrelated young horses. They performed their study on forty two 1 to 3 year old French Saddlebreds all raised in the same farm and with similar housing and handling conditons since birth. In a first test, inspired by classical open-field tests but carried out in a familiar environment so as to evidence social aspects (Jones, 1977), the horses were alone in an arena (arena test) and their behaviour was recorded by scans made at 10 second intervals (posture, activity) whereas rare behaviours like rolling or defecation were recorded every time they occurred. In the second test, an unknown object (a grid with red ribbons) was present in the arena. The third test consisted of leading the horse over a 'bridge' made of wooden planks (bridge test) and the time required to cross was measured.

For all three tests, these authors found that half siblings showed more similar behaviours than unrelated horses, and full siblings showed even more similarities. Lankin and Bouissou (1998) observed the same phenomenon in a study performed on 82 female Welsh ponies.

The same tests performed on a large sample of horses revealed differences between breeds, especially in the bridge test where purebred Arabs and Thoroughbreds required the most time in order to cross and Fjord ponies, Haflinger or Quarter horses the least time (Hausberger *et al.*, 1998; Hausberger *et al.*, 2004). In overall, Fjord, Merens, New Forest or Camargue horses appeared to be the least reactive to anxiogenous situations.

Reactions to humans: Kusunose (in Houpt and Kusunose, 2001) evaluated the reactions of 9644 young thoroughbreds sired by 62 stallions in two training centers during veterinary visits. A grade was given for each horse for ease of handling. It appeared that animals descending from a given stallion tended to behave similarly whether or not in the same training center. This fortifies the evidence of a paternal effect on behaviour.

Hausberger and Muller (2002) adopted the same type of evaluation to test reactions of adult horses to unknown approaching humans. Two hundred and twenty four horses (Angloarabs, French Saddlebred, Thoroughbreds) stabled at the Saumur 'Ecole Nationale d'Equitation' were tested. All were geldings between 4 and 15 years old and living under similar conditions. Differences among breeds emerged: the thoroughbreds were indifferent more often than Angloarabs and French Saddlebreds; French Saddlebreds showed positive attitudes more often than Angloarabs.

Questionnaires, surveys, reports

Questionnaires have been used in a variety of studies, some aiming at characterising the horses' personality (Mills 1998, Morris *et al.*, 2002), others aiming at giving an overall view of different breeds, as did Mc Greevy *et al.* (1995) and Luescher *et al.* (1998) for stereotypies (see part I). An overview of breeds' temperament appeared in Hayes (1998) based on reports from 50 trainers or veterinarians about 10 breeds and 6 traits (training aptitude, will to work, fear ...). Differences reported between breeds were for example skittiness, present more frequently in Purebred Arabs and Thoroughbreds than in Appaloosas, while Thoroughbreds were reported to lack concentration in training more than other horses.

Evidence of an influence of environmental factors on horses' behaviour

A large variety of environmental factors may influence the horses' behaviour. Amongst them, it has been shown that housing conditions in particular like feeding roughage, type of bedding, visual access may all be involved in the emergence of stereotypic behaviours (e.g. Mc Greevy *et al.*, 1995). Impact of the general environment may include stabling versus paddock, group versus isolated housing, type of food (e.g. Holland *et al.*, 1996).

Other environmental factors involve the influence of conspecifics or of the human-horse relationship. Thus, the mother's social ranks influence the foals' ranks (Houpt and Wolski, 1980). Different studies have reported an influence of human-horse relationship on the horses' behaviour. Amongst the earlier studies, Heird *et al.* (1986) revealed an impact of handling foals on subsequent learning abilities.

More intriguing is the finding of relations between the type of work and behavioural aspects like stereotypies, emotional reactions or learning abilities. Thus both Mc Greevy *et al.* (1995) and Hausberger *et al.* (1996) found that more dressage horses tended to express stereotypic behaviours than horses performing other types of work. This finding is reinforced by the latter study, where gelding horses, observed in their box, were all housed in the same general conditions.

Experimental tests performed on 120 geldings at the Saumur 'Ecole Nationale d'Equitation' (Hausberger *et al.*, 1996) and on more than 700 horses in a large scale study (Hausberger *et al.*, 2004; Hausberger and Richard-Yris, 2005) confirmed a higher general emotional state of dressage, especially high school horses, as compared to other show horses, but also a higher general emotional state of show horses as compared to leisure horses.

Finally, the identity of the usual caretaker appeared as a major factor determining the reactions of adult gelding horses to the sudden approach of an unknown person in a study performed on 224 horses (Hausberger and Muller, 2002).

These few examples illustrate the large variety of possible environmental influences which necessarily interact or add with genetic bases in determining the horses' behaviour.

The interplay between genetic and environmental factors

Few studies have investigated precisely how both influences may act on the horses' behaviour. In a study made of over 700 horses belonging to 16 different breeds and located in more than 100 different sites, Hausberger *et al.* (1998; 2004) investigated the effects of breed, sire and of different environmental factors on both emotional reactions and learning abilities.

Nine factors were investigated: sire, breed, site, type of work, housing (box/paddock), food, number of riders, sex, age. Multivariate analyses (FCA, AHC) and analyses of variance revealed that seven of these factors had a significant effect (excluding sex and age) and three had a strong weight on variance: sire, breed and site. Another environmental factor, the type of work, appeared as most important (see also part II). The relative weight of the different factors appeared to depend on the behavioural trait tested, the genetic factors prevailing for neophobic reactions (novel object, bridge test) and environmental factors (type of work, site) on learning or reactions to social isolation.

Hausberger *et al.* (1998; 2004) compared the reactions of 98 breeding stallions living in similar conditions but in different studs to experimental tests. They found the same breed effects in all studs where in particular purebred Arabs appeared more emotional than Angloarabs and both more emotional than French Saddlebreds. On another hand, large differences appeared between studs, some showing a majority of calm horses as compared to others where all horses tended to react strongly to the tests. As a result, a purebred Arab at one stud could in fact be less reactive than a French Saddlebred at another one. In the present case, genetic (breed) and environmental (stud) factors acted additively.

Interference between both sorts of influences are probably especially involved in the expression of stereotypic behaviours. An illustration is given by Houpt and Kusunose (2001): none of the foals living with a Przewalski mare, who cribbed with a high frequency, expressed this behaviour. However when one of her colts was sent away as an adult, he began cribbing in the new place. Environmental stress (social separation, confinement ....) may elicit the expression of abnormal behaviours in horses with genetical susceptibility.

Can horses be selected on the basis of behaviour ?

All these review leads one to think that there is a mixture of environmental and genetic effects on the behaviour of the horse. This mixture produces differences between breeds and between horses of the same breed but with different parents. In order to use this genetic variability in selection, a precise measure of the influence of genetic and environmental effects is needed. That is, we need to estimate heritability of typical traits of behaviour and not just test the significance of some differences between breeds.

Such estimates are available for the notation used in the stallion test to judge the 'temperament' of the horse in Germany and in Sweden. Heritability of this subjective measure is between 0.24 to 0.41 in Germany (Ricard *et al.*, 2000) and 0.33 in Sweden (Gerber-Olsson *et al.*, 2000). Obviously, it is not a true measure or test of behaviour but a notation of the use of the horse in the context of sport competition.

The very first estimates of heritability on experimental tests to evaluate individual behavioural tendencies are now available in the study of Hausberger *et al.* (2004, see above). The analysis will be developed here as an example of what could be determined about heritabilities.

Material and method

The tests have been built by the team of Hausberger on 703 horses all over France, form 16 different breeds. Five scores are reported: notation of emotionality for a horse alone in a known place (A), alone with a strange object (O), time spent to cross a strange 'bridge' (H), time spent to open a wooden box (B1) and the repetition of this test to evaluate learning (B2). The distributions of (A) and (O) are clearly not normal, with a high dissymmetry, like a normal distribution cut in the middle. The variables (H), (B1), (B2) have the disadvantage that the maximum time recorded score was given to 30% to 40% of the horses tested. So, for the measurements (A) and (O), a Box-Cox transformation was used in order to normalise the data. The transformation is:

$$z = \frac{y^t - 1}{t\dot{y}^{t-1}} \quad \text{with } \dot{y} = \left(\prod_{i=1}^{n} y_i \right)^{1/n} \qquad (1)$$

The value is chosen to maximise the criteria of normality of the distribution (test de SHARIPO-WILK), t is equal to 0.265 for (A) and 0.130 for (O). For the variable (H), (B1) and (B2) a probit transformation was necessary: the horses with the time maximum are allocated the deviation of normal distribution corresponding to the percentage of higher tail of normal distribution.

The model used is the following:

Measure = site + age + sex + breed + type of work + number of riders + Housing + food + breeding value of the horse + residual.

With:
Effect of site: 106 sites where identified, sites with less than 10 horses were grouped, so 15 different levels remain.

Effect of age: horses are aged 2 to 20 years, 14 levels were defined: from 2 to 15 years and more.
Effect of breed: 16 breeds were tested.
Effect of type of work: 13 levels were defined: team (10), eventing (17), riding school (121), jumping (146), dressage (46), different riding school (18), reproduction for stallion (117), broodmare (35), foal (60), show (26), trekking (71), voltige (6), western(30)
Effect of the number of riders: 5 levels were defined: any, high level instruction, riding school, middle level instruction, owner.
Effect of housing: box or paddock
Effect of food: traditional, industrial food, mixed.
Effect of breeding value of the horse: All horses have a different breeding value which is correlated to the breeding value of their relatives. All relationships were taken into account in the model, not only relationship between half-sibs from the same sire. 4880 ancestors were extracted from the pedigree file managed by SIRE (Système d'Idenficiation Répertoriant les Equidés). So it is the relationship between 5583 (4480+703) horses which were included in the analysis. It is the ratio of the variance estimated for this effect and the total variance (genetic variance and residual variance) which give the heritability of the trait.

Heritability is not supposed to be the same in different breeds, but the number of horses measured by breed is not sufficient to allow the estimation of different heritabilies in each breed; therefore, it was estimated on the whole data, and an example of what it could be is given for the larger breed in the sample, the Selle Français.

Heritability is estimated with a REML (Restricted Maximum Likelihood) using the software VCE (Groeneveld, 1997).

Results

Phenotypic correlation

The correlations after transformation of the variables in order to normalised then are higher (in absolute value) than on the original variables. There are clearly 3 types of test (A) and (O), (H), (B1) and (B2). The 3 traits displayed have low correlations: (A) and (O) with (H) is 0.02 and 0.14, (A) and (O) with (B) is -0.04 to 0.02, (H) with (B) is 0.27 to 0.31.

Table 1. Correlation between measurement in 5 temperament tests (transformed variables).

		A	O	H	B1	B2
Alone in an area	A		0.516	0.023	-0.042	-0.004
Novel Object	O	0.516		0.144	-0.041	0.017
Handling travel	H	0.023	0.144		0.311	0.269
Opening box	B1	-0.042	-0.041	0.311		0.920
Memorisation	B2	-0.004	0.017	0.269	0.920	

Effect of genetic

The breed effect is significant for (A), (O), (H). Results are in Table 2.

Results must be taken with caution according to the number of horses per breed and the possibility of non random sampling, but results are interesting especially for Selle Français, Anglo-Arab, Arab, and Merens.

Table 2. Breed effects (in standard deviation unit).

Number of horses	*Breed*	Test*: estimate, (confidence interval)			
		A	O	H	B1
21	Thoroughbred	-0.52 (0.25)	-0.29 (0.19)	0.35 (0.18)	0.33 (0.22)
23	French Trotter	-0.42 (0.29)	-0.53 (0.22)	-0.25 (0.20)	0.12 (0.25)
76	Anglo-Arab	-0.32 (0.13)	-0.03 (0.10)	0.11 (0.09)	0.01 (0.12)
246	Selle Français	0.00 (0.12)	0.00 (0.09)	0.00 (0.09)	0.00 (0.09)
41	Arab	-0.32 (0.22)	0.22 (0.16)	0.65 (0.14)	0.26 (0.19)
26	Camargue	-0.51 (0.27)	-0.48 (0.21)	-0.27 (0.20)	-0.06 (0.24)
29	Spanish Horse	-0.91 (0.32)	-0.49 (0.25)	-0.15 (0.23)	-0.34 (0.28)
20	Appaloosa	-0.33 (0.41)	-0.01 (0.32)	-0.43 (0.30)	-0.24 (0.37)
29	Quarter Horse	-0.37 (0.65)	0.06 (0.52)	-0.82 (0.48)	-0.55 (0.58)
24	Connemara	-0.26 (0.32)	-0.12 (0.23)	0.36 (0.21)	-0.29 (0.27)
14	Fjord	-0.79 (0.41)	-0.64 (0.32)	-0.91 (0.30)	0.04 (0.37)
28	Haflinger	-0.09 (0.27)	0.02 (0.20)	-0.46 (0.18)	-0.57 (0.24)
63	Mérens	-0.62 (0.35)	-0.68 (0.26)	-0.20 (0.23)	-0.03 (0.31)
25	New-Forest	-0.67 (0.27)	-0.58 (0.21)	-0.29 (0.19)	0.17 (0.24)
11	French Pony	-0.39 (0.54)	0.15 (0.41)	-0.27 (0.38)	-0.36 (0.48)
22	Welsh	-0.71 (0.30)	-0.20 (0.23)	-0.01 (0.21)	0.11 (0.27)

*: A=Alone in an area, O=Novel Object, H=Handling travel, B1 =Opening box, B2=Memorisation

The breed which reach the higher level for (A), therefore the more emotional breed, are the Selle Français and the Halfinger and Connemara. The less emotional are the New Forest, the Welsh and the Spanish horse. Comparable results are obtained for the correlated trait (O), except for the Arab which become the more emotional. The longer times spent to pass the bridge are for the Arab horses, the thoroughbred and the Connemara. The shorter times are for the Quarter Horse and the Fjord. So the profile is different according to the test. The understanding of behaviour needs thought because even if the more sporting breeds seem the more emotional (ex: Selle Français), and less efficient for the handling test (ex: thoroughbred), this is not systematic (the Arab horse is also bad for handing and the thoroughbred is not emotional in the alone test).

Results of heritability are in Table 3. The accuracy of estimates is low. The confident interval of heritability for A and B1 is from 0.20 to 0.60, from 0.10 to 0.50 for O and H and from 0 to 0.40 for B2. But an heritability estimated on a multitude of breed is few efficient for a selection in a particular breed. That is why results for the Selle français only are given, even if confidential interval are very high and strange estimates appears (0.80) suggesting a problem of sampling. Only 18 sires have more than 4 offspring and those sires are already related (sire and son). However, theses estimates are the very first one on horses and it is important to note that they seem rather high and so it would be particularly interesting to pursue the research.

Table3. Heritability of transformed measure of temperament tests (standard deviation).

	A	O	B1	B2	H
All breeds	0.42 (0.11)	0.29 (0.12)	0.36 (0.13)	0.17 (0.14)	0.24 (0.10)
Selle Français	0.39 (0.26)	0.40 (0.24)	0.48 (0.20)	0.28 (0.21)	0.81 (0.17)

Toward a selection

All the research on genetic of behaviour leads to manage a place for the influence of gene. In a practical point of view, research much now be done to quantify this relation and to integrate such trait into a selection plan. But a lot of questions remain to be answered I order to reach this objective. The first point is the estimation of heritability. To have a correct estimation of heritability, we need at least 1000 horses measured from no more than 25 sires. The knowledge of heritability guides the selection into different strategies: a high one privileges a selection on own performance, even reserved to stallions. A low one leads to a selection on progeny or a marker assisted selection. The second point is the correlation between the different traits recommended on the basis of behaviour and the major selection objective of the breed in order to at least maintain the selection on the sport trait involved. The third point is to choose a good measure for the trait to be selected: few studies have been done to compare physiological traits, test traits and judgments based on human opinion. Visser *et al.* (2003) didn't find any correlation between tests for behaviour and heart rate on 42 horses but they found a phenotypic correlation between these tests and notation of judge under riders at 3 years old. The relation between the test and the objective must be defined.

In all cases, nothing is neutral. At present, sport horses are selected for jumping competition for example and so, a special behaviour is certainly selected simultaneoustly. A mixture between reactivity and sweet behaviour in the so special context of a show jumping.

Conclusion: being aware of genetic predispositions in order to increase welfare

All the evidence presented here strongly suggests that genetic bases are indeed involved in the horses' behaviour. Environmental factors, at all stages of development, will help developing the individual personality as a result of additive or interactive effects with genetic predispositions. The example of stereotypic behaviours is typical in that it is more a susceptibility to environmental constraints than a behaviour per se which is being transmitted.

Obviously, horses issued from different breeds or sires do not react the same way to similar conditions. Being aware of such underlying tendencies may be of great help in adjusting working or housing conditions or choosing a horse that will fit with the conditions one can offer.

Selection on the basis of individual behavioural characteristics is obviously a possibility in a near future, but no selection will be able to palliate deficiencies in the horses' environment, amongst which the human-horse relationship appears as a major component.

References

Bates, J.E., 1989. Concepts and measures of temperament. In: Temperament in childhood. G.A. Kohnstamm, J.E. Bates and M.K. Rothbart (editors),. Wiley, Chichester, 3-26.

Bateson, P., 2000. Behavioural development in the cat. In: The domestic cat. The biology of its behaviour. 2nd edition., D.C. Turner and P. Bateson (editors), Cambridge University Press, Cambridge, 9-22.

Belyaev, D.K. and L.N. Trut, 1975. Some genetic and endocrine effects of selection for domestication in silver foxes. In: The wild canids, M.W. Fox (editor), Van Nostrand Reinhold, New York, 416-426.

Bruns, E., 1982. Entwicklung der Leistungsprüfung und Datennutzung für Reitpferdezucht. Züchtungskunde 54: 446-453.

Budzynski, M., L. Soltys and J. Wawiorko, 1992. Estimate of excitability of half-bred horses. In: 43rd Annual Meeting of the EAAP.

Crowell-Davis, S.L., 1986. Spatial relations between mares and foals of the Welsh pony. Animal Behaviour 34: 107-1005.

Dodman, N.H., J.A. Normile, L. Shuster and W. Rand, 1994. Equine self-mutilation syndrome (57 cases). Journal of the American Veterinary Medical Association 204: 1219-1223.

Faure, J.M. and A.D. Mills, 1998. Improving the adaptability of animals by selection. In: Genetics and the behavior of domestic animals, T. Grandin (editor), Academic Press, New York, 235-264.

Fordyce, G., M.E. Goddard and G.W. Seifert, 1982. The measurement of temperament in cattle and the effect of experience and genotype. Proceedings of the Australian Society of Animal Production 14: 329-332.

Gerber Olsson, E., T. Arnason, A. Nasholm and J. Philipsson, 2000. Genetic parameters for traits at performance test of stallions and correlations with traits at progeny tests in Swedish warmblood horses. Livest. Prod. Sci. 65: 81-89.

Gerlai, R. and V. Csañyi, 1990. Genotype-environment interaction and the correlation structure of behavioral elements in paradise fish (*Macropodus opercularis*). Physiology and Behavior 47: 343-356.

Goddard, M.E. and R.G. Beilharz, 1982. Genetics of traits which determine the suitability of dogs as guide-dogs for the blind. Applied Animal Ethology 9: 299-315.

Groeneveld E., 1997. VCE4 User's guide and reference Manual. Federal Agricultural Research Center, Neustadt, Germany, 49pp.

Hall, C.S., 1941. Temperament: a survey of animal studies. Psychol. Bull. 38: 909-943.

Hausberger, M., C. Bruderer, N. Le Scolan and J.S. Pierre, 2004. Interplay between environmental and genetic factors in temperament/personality traits in horses (*Equus caballus*). J. Comp. Psychol., 118(4): 434-446.

Hausberger, M., N. Le Scolan, C. Bruderer and J.S. Pierre, 1998. Le tempérament du cheval: facteurs en jeu et implications pratiques. In: 24e journées d'étude de la Recherche Equine, 4 mars 1008, Paris, Institut du Cheval, 159-169.

Hausberger, M., N. Le Scolan, C. Muller, E. Gautier and A. Wolff, 1996. Caractéristiques individuelles dans le comportement du cheval: prédictibilité, facteurs endogènes et environnementaux. In: 22ème Journée d'étude, Institut du Cheval, 113-123.

Hausberger, M. and C. Muller, 2002. A brief note on some possible factors involved in the reactions of horses to humans. Appl. Anim. Behav. Sci. 76: 339-344.

Hausberger, M. and M.A. Richard-Yris, 2005. Selection and genetics of behaviour. In: The domestic horse, D. Mills and S. Mc Donnell (editors), Cambridge University Press, Cambridge, in press.

Hayes, K.E.N., 1998. Temperament tip-offs. Horse & Rider 37: 46-51.

Heird, J.C., D.D. Whitaker, R.W. Bell, C.B. Ramsey and C.E. Lokey, 1986. The effects of handling at different ages on the subsequent learning ability of 2-year-old horses. Applied Animal Behaviour Science 15: 15-25.

Holland, J.L., T.N. Meacham and D.S. Kronfeld, 1996. Behavior of horses is affected by soy lecithin and corn oil in the diet. J. Anim. Sciences 74: 1252-1255.

Hosoda, T., 1950. On the heredity of susceptibility to windsucking in horses. Japanese Journal of Zootechnical Science 21: 25-28.

Houpt, K.A. and R. Kusunose, 2001. Genetics of behaviour. In: The genetics of the horse. CABI Publishers.

Houpt, K.A. and T.R. Wolski, 1980. Stability of equine hierarchies and the prevention of dominance related aggression. Equine Vet. J. 12: 18-24.

Houpt, K.A., M.S. Parsons and H.F. Hintz, 1982. Learning ability of orphan foals, normal foals and of their mothers. J. Animal. Sci. 55: 1027-1032.

Jones, R.B., 1977. Sex and strain differences in the open-field responses of the domestic chick. Appl. Anim. Ethol. 3: 255-261.

Juarbe-Diaz, S.V., K.A. Houpt and R. Kusunose, 1998. Prevalence and characteristics of foal rejection in Arabian mares. Equine Veterinary Journal 30: 424-428.

Langlois, B., 1984. Cheval de loisir et de sport: aptitudes et sélection. In: Le cheval. Reproduction, sélection, alimentation, exploitation, R. Jarrige and W. Martin-Rosset (editors), Editions INRA, Paris, 423-435.

Lankin and M.F. Bouissou, 1998.- Interspecific differences in fear of humans ('domestic behaviour') in farm animals. In: Proceedings 32nd Congress of the International Society for Applied Ethology, 114.

Le Neindre, P., G. Trillat, J. Sapa, F. Ménissier, J.N. Bonnet and J.M. Chupin, 1995. Individual differences in docility in Limousin cattle. Journal of Animal Science 73: 2249-2253.

Luescher, V.A., D.B. Mc Keawn and H. Dean, 1998. A cross sectional study on compulsive behaviour in horses. Equine Vet. J. 27: 14-18.

Mader, D.R. and E.O. Price, 1980. Discrimination learning in horses: effects of breed, age and social dominance. Journal of Animal Science 50: 962-965.

Marinier, S.L. and A.J.Alexander, 1992. Use of field observations to measure individual grazing ability in horses. Animal Sciences 66: 475-481.

McCune, S., 1995. The impact of paternity and early socialisation on the development of cats' behaviour to people and novel objects. Applied Animal Behaviour Science 45: 109-124.

McGreevy, P.D., N.P. French and F.J. Nicol, 1995. The prevalence of abnormal behaviours in dressage, eventing and endurance horses in relation to stabling. Vet. Record 137: 36-37.

Mills, D., 1998. Personality and individual differences in the horse, their significance, use and measurement. Equine Veterinary Journal Suppl 27: 10-13.

Morris, P. H., A. Gale and K. Duffy, 2002. Can judges agree on the personality of horses? Personality and Individual Differences 33: 67-81.

Ricard, A., 1997. Les nouveaux indices de sélection. Institut du cheval.

Ricard, A., E. Bruns and E.P. Cunningham, 2000. Genetics of performance traits. In: The genetics of the horse. CABI Publishers., 281-306.

Redbo, I., P. Redbo-Torstensson, F.O. Odberg, A. Hedendahl and J. Holm, 1998. Factors affecting behavioural disturbances in race-horses. Animal Science 66: 475-481.

Tyler, 1972. The behaviour and social organization of the New Forest ponies. Anim. Behav. Monogr. 5: 85-196

Visser, E.K., C.G. Van Reenen, M. Rundgren, M. Zetterqvist and K. Morgan, 2003. Responses of horses in behavioural tests correlate with temperament assessed by riders. Equine Veterinary Journal 35: 176-183.

Vecchioti, G.G. and R. Galanti, 1986. Evidence of heredity of cribbing, weaving and stall-walking in thoroughbred horses. Livestock Prod. Science 14: 91-95.

Waring, G.H., 1983. Horse behavior. The behavioral traits and adaptations of domestic and wild horses, including ponies. Noyes Publications, Park Ridge, NJ.

Waring, G.H., 2003. Horse behavior. 2nd edition. Noyes Publications/William Andrew Publishing, Norwitch, NY.

Wolff, A. and M. Hausberger, 1992. Comparaison de caractéristiques comportementales chez des poulains: une étude quantitative. In: CEREOPA 1992, 18ème Journée d'Etude, 78-90.

Wolff, A. and M. Hausberger, 1994. Behaviour of foals before weaning may have some genetic basis. Ethology, 96: 1-10.

Wolff, A. and M. Hausberger, 1996. Learning and memorisation of two different tasks in horses: the effects of age, sex and sire. Appl. Anim. Behav. Sci., 46: 137-143.

Wolff, A., M. Hausberger and N. Le Scolan, 1997. Experimental tests to assess emotivity in horses. Behav. Processes, 40: 209-221.

Part B

Early experience, human-foal relation and the impact of training

Early experience and behavioural development in foals

M. Hausberger, S. Henry and M.A. Richard-Yris
Université Rennes 1, UMR 6552 Ethologie - Evolution - Ecologie, Campus de Beaulieu, F - 35042 Rennes Cedex, France

Abstract

The young age is an important time period for the behavioural development in animals. Different experience factors can influence this development, including housing, social life and human foal interactions. Published studies indicate that direct and intensive handling of foals has rarely long term beneficial effects. Weaning may be more favourable for such handling. On another hand, the comparison of different approaches suggests that indirect approaches through the mother as social model may be a fruitful practice with long term effects.

Keywords: development, management practices, social facilitation, foal handling

Introduction

The development of individuals begins at birth, or even before and ends at its death (Magnusson and Cairns 1995). However, it is generally admitted that the young age is an essential step in individual construct. It is a time when experiences can have an important impact, given the characteristics of young animals: openness to learning, 'curiosity' for novel stimulations … Numerous factors are involved in this individual behavioural development: physical (box, paddock) or social mother and then social group) environments. In domestic animals, it has been shown that the relation to humans is part of these multiple factors. Thus, short periods of positive handling can lower the stress related to a human presence and lessen its negative effect on welfare (Dawkins, 1980). Pigs that are stroken at each spontaneous approach of human (gentling) increase their frequencies of approach (Hemsworth and Barnett, 1992). In cattle, early handled animals are quieter in a human presence (Sato *et al.*, 1984; Boissy and Bouissou, 1988). A positive impact of early contacts on later responses of animals to humans has been demonstrated in numerous species including dogs (Scott and Fuller 1965; Wright, 1983), chickens (Murphy and Duncan, 1978; Jones and Faure, 1981), silver foxes (Belyaev and Trut, 1975; Fordyce *et al.*, 1985) and rabbits (Kersten *et al.*, 1989). These numerous studies have revealed three key factors in the establishment of human - animal relation (genetic predispositions): (1) the nature, quality and frequency of handling (Hemsworth *et al.*, 1992; Jones, 1994), (2) the time period when handling occurs (Hediger, 1968), (3) the social environment (Nicol, 1995).

Moreover, the individual itself is an actor of its development: its own temperamental traits (related to some part to genetic bases) have an impact on degree of influence of experiential factors, such as the establishment of social relations (Hausberger *et al.*, 2004; Hausberger and Richard-Yris, 2005). There is at present few information on these aspects in young horses.

Breeding practices and temperament

Few studies have considered the influence of the way of life on the behavioural development of young horses. One can suppose that a variety of environmental factors are involved in adults: housing in box, social environment, amount roughage (Mc Greevy *et al.* 1995a), relation to human via training (Mc Greevy *et al.* 1995b, Hausberger *et al.* 1996) or through daily care (Hausberger and Muller 2002).

The few studies performed on housing conditions converge towards on importance of social experience (Sondergaard and Shougaard 2000, Christensen *et al.* 2002). Thus, Heleski *et al.* (2002) have compared the behaviour just after weaning of foals kept singly in a box to those of foals kept as a group in a paddock. Important differences appeared: the animals kept in groups had a larger behavioural repertoire. Moreover, all the foals kept singly in a box showed at least one abnormal behaviour, from pawing repeatedly to lick frequently the boxe's walls. These observations were confirmed by Rivera *et al.* (2002) who showed also that young horses kept in groups in a paddock adapted more easily to training than the singly kept animals. It is however difficult in these studies to separate the effects of the social and physical (box / paddock) environment.

In a recent study, Sondergaard and Halekoh (2003) show that one and two year old horses kept singly in box seeked more human contact than group-raised animals. However these differences diminish with time, possibly through the daily contact with caretakers when they bring food in winter. Similarly, young horses that were handled three times a week for 2 months were approached more easily than non handled animals at the age of 1 year but again this difference disappears in the following year. The presence of social partners may therefore enhance (Rivera *et al.* 2002) or restrict (Sondergaard and Halekoh 2003) the human - horse relation.

In a epidemiological study performed on 21 breeding farms in Brittany, 170 young horses, aged between 1 and 3, were tested while the breeding practices were carefully noted (way of life, handling: how, when ...). Large differences appeared between farms in the reaction of yearlings to humans: in some farms, all horses approached spontaneously, in others, none approached or even obviously fled when a human entered the paddock. Experimental tests performed on the 2 and 3 year old horses in the same farms revealed similarly a gradient between farms where the animals were not emotive and learned easily an instrumental task and those where the animals showed strong emotional reactions in novel situations and learned badly. Interestingly, the same farms showed calm 2 / 3 year old horses and familiar yearlings.

When these results are confronted to practices, it appears that the 'best' farms correspond to those where handling occurs mainly around weaning and / or in the following year (SF7, SF4) while the farms showing 'bad scores' were characterised by either very intensive and permanent handling (from imprinting to daily halter fitting, handling, leading ...) (SF9) or by no handling at all after weaning and the following year (T4, SF1, T9) (Figure 1).

Therefore there is no simple relation between the amount of handling and the quality of the horse - human relation or the emotional state of the animal. 'Excessive' handling may well bring aversive responses (Hausberger and Richard-Yris, 2003).

Should the newborn foal be 'imprinted'?

Whereas the idea of 'imprinted of newborn foals' proposed by Miller (1991) has been readily accepted in a number of breeding farms, this is much less the case in the scientific world, where few publications to date have studied its effects.

There are several reasons for that:

- Imprinting as defined by Lorenz (1935), on the basis of observations of nidifugous birds consists in 'acquiring an object' towards which the 'instinctive social reactions' will be oriented. Criteria are very precise: existence of a critical period, irreversible phenomenon... This mechanism is considered as a basis for future sexual choices!
- Imprinting refers to the young animal following a mobile object. This does not infer habituation, desensitisation or any other specific aspect.

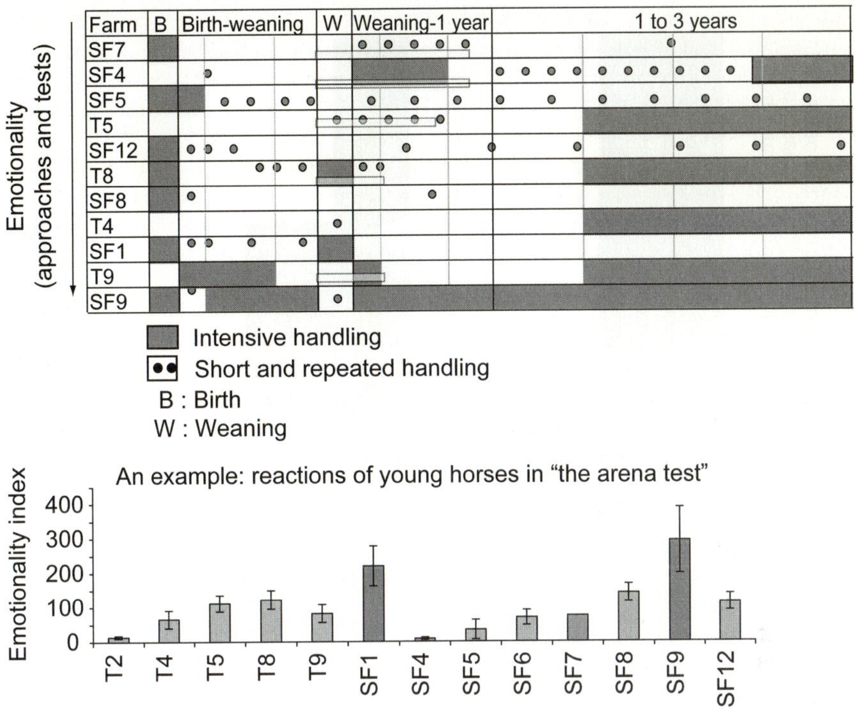

Figure 1. Farm practices and emotional reactions of young horses.

- Imprinting is not related to restricting movements, which on the contrary induces a stress. Different authors have suggested that the apparent decontraction of the foal at the end of such handling corresponds rather to learned helplessness, which has known negative effects (Seligman *et al.* 1968, Simpson 2002, Sigurjondottir pers. com.)

It remains to be known whether this early stress is justified in terms of later ease of handling, habituation to stimuli or lowered emotional reactions.

The two main studies performed give some contradictory results. Williams *et al.* (2002) who handled the foals 2, 12, 24 and 48 h after birth found no beneficial effect on the behaviour of the foals when 1 and 4 months old (Figure 2).

The same approach performed between 1 and 5 days by Simpson (2002) showed that handled foals approached a familiar person (but not a stranger) more easily at the age of 3 months. The handling did not show any positive effect on the reactions of foals towards specific stimuli. The imprinted foals reacted as strongly as the control foals to halter fitting or shearing at 3 months. A good part of the specific stimulations (spray, plastic bag …) used for habituation while imprinting revealed no lasting effect. According to Simpson (2002), the slight positive effect on approachability might well be obtained as well with later handling.

Two other unpublished studies (Diehl *et al.*, Sigurjondottir and Gunnarson, pers. com., 2002) underline the important resistance of the foals to the procedure. The second study, performed on Icelandic horses, indicates a correlation between the foal's resistance and the dam's temperament. In both cases, the imprinted foals, that were not regularly handled afterwards, were as difficult to

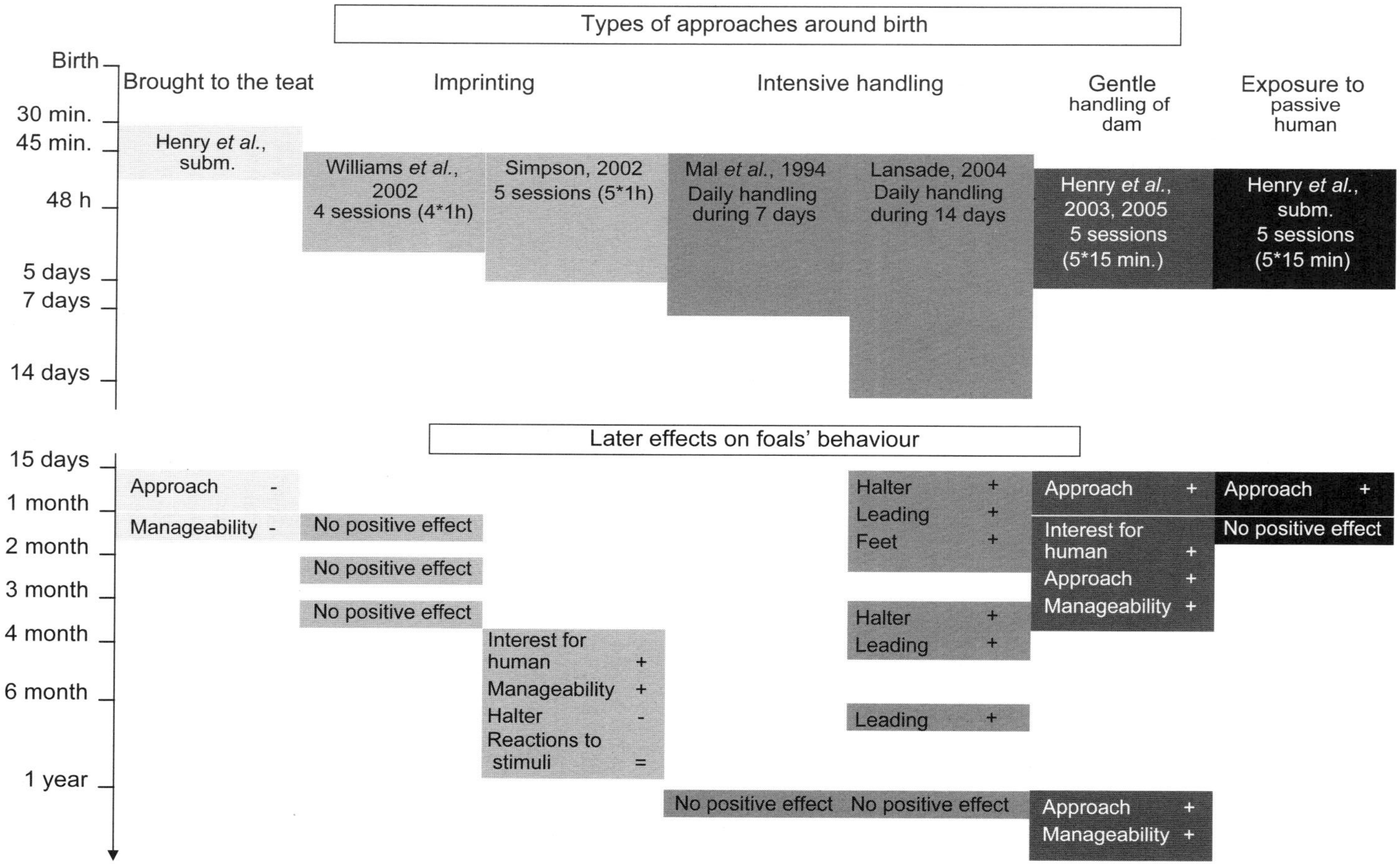

Figure 2. A general view of the findings from approaches at birth (+: positive effect, -: negative effect, =: no difference when compared to controls).

approach as control foals at the age of 4 months. Therefore the question of the durability of the effects is important.

Does early handling have lasting effects?

Most studies bring the conclusion that early handling does not have long lasting positive effects. In a pioneering study, Waring (1983) compared the reactions of foals either intensively handled, confronted to the passive presence of a human or to a dummy between 5' and 15 h post partum. At the age of 2 days, the handled foals accepted more easily a halter but at the age of 90 days, no difference could be observed. In Mal *et al.* (1994)'s study, foals handled during 7 days after birth with or without session per week until weaning were not easier to handle 120 days after weaning than non-handled foals. The same type of handling (halter, stroking …) performed during 14 days after weaning gave short term effects that diminished with time: at 6 months of age, handled and control foals did not differ (Lansade *et al.*, 2004).

At short term, less intrusive handling than imprinting seems to have positive effects on halter fitting, and leading and emotional relation to humans (Heird *et al.* 1986, Mal and Mc Call 1996, Jezierski *et al.* 1999, Lansade *et al.* 2004).

Are there better time periods for establish the human - foal relation?

Whereas the 'imprinting' supposes the existence of a 'critical period' at birth, we have little information on the existence of favourable periods for foal handling, while more is known on puppies or kittens (Karsh and Turner 1988, Scott and Fuller 1965). In cattle, weaning seems to be a good time period for handling (Boissy and Bouissou 1988, Le Neindre *et al.* 1994).

In horses, handling at weaning may increase the learning abilities (Heird *et al.*, 1986). Repeated handling in the 12 days after weaning have long lasting effects on the foal's maniability during at least 18 months. On the contrary, foals handled between the 21st and 33rd days after weaning were not easier to handle after 12 months (Lansade *et al.*, 2004). At this particular period when foals are socially deprived, they may be more open to human contact. Some breeding farms appeared indeed to choose this time period for handling (see above). However, young animals remain 'open' to later approaches as the results of Sondergaard and Halekoh (2003) suggest. The 2 year old non handled animals become as familiar as the handled animals probably through the mere daily exposure to the caretakers bringing food. Therefore, the most promising research line may well be in the way the relation is established rather than the moment.

Towards other possible approaches

In the whole, all these results tend to indicate that very intensive handling has rarely long lasting positive effects and might even have some negative effects. On another hand, a mere exposure to humans or the dam's reactions seem to influence the foal's responses. Therefore the question arises of whether less intrusive approaches could have positive long lasting effects.

We have performed a study aiming to comparing the importance of different approaches at birth or in the following days on the human - foal relation. Three approaches were tested:

- A mere daily exposure during 15 mn to a passive human in the box in the five days after birth.
- A gentle handling of the mare (soft brushing and food reward) in the front of the foal, without any direct attention from the experimenter to the foal. The mother is a familiar and secure element that is known to help the young individual to explore the environment (Ainsworth *et al.* 1978) and

in other species the mother's reactions to humans may be transmitted through social facilitation (Lyons *et al.* 1988; Ruiz-Miranda and Callard, 1992; Bertin and Richard-Yris, 2004).

- We also tested the effects of a common practice in breeding farms that consists in bringing the foal to the dam's teat. This is a very short handling that was here performed 30' after birth.

When the effects of the 3 types of handling were compared at later ages, it appeared clearly that the best results are obtained via the dam (Figure 3). The foals that have been brought to the teat did not avoid a mere approach but refused any direct contact. Moreover they did not show any habituation, contrarily to the controls.

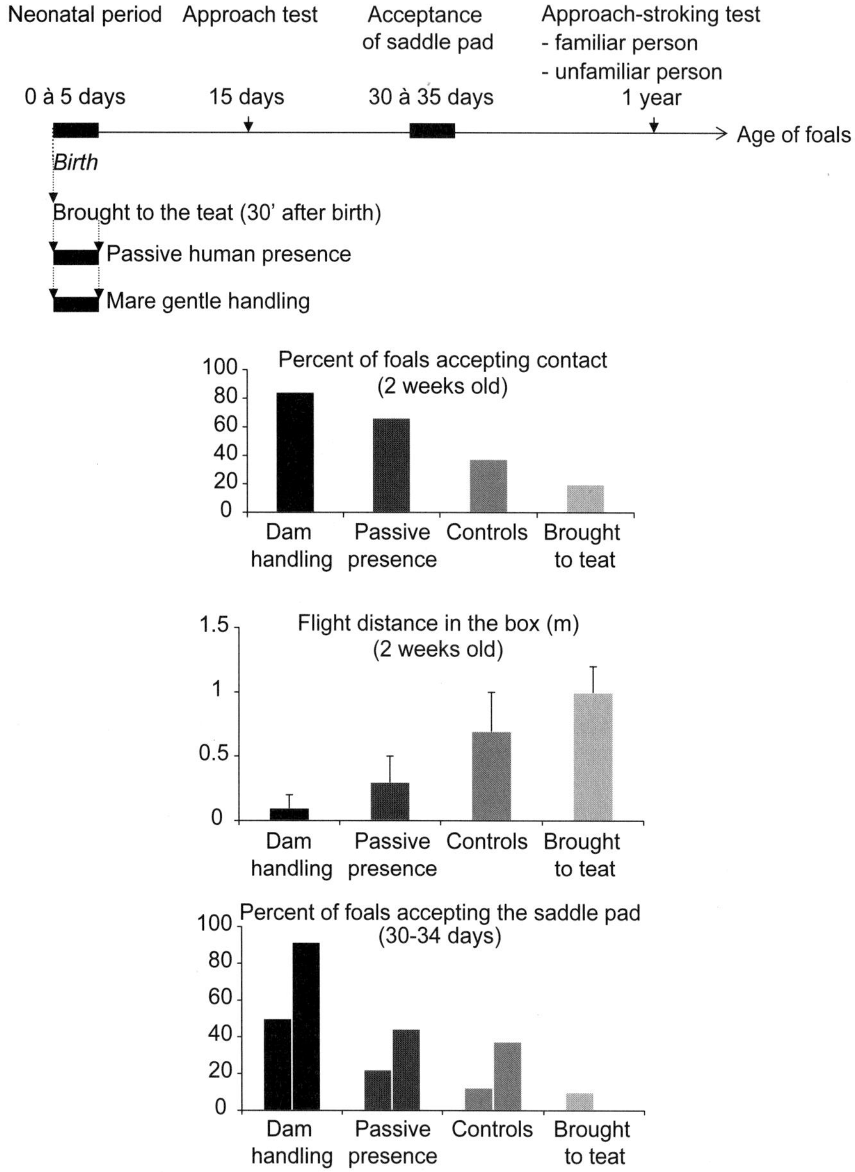

Figure 3. Ease of approach of foals at 15, 30, 34 days of age according to the experience with humans at birth or in the following days.

The foals exposed to a passive human presence showed lower flight reactions than the controls at the age of 2 and 4 weeks but the effects were not long lasting. The foals whose dams had been handled approached very readily the experimenter at 2 weeks, accepted easily a direct contact and accepted for most of them to receive a saddle pad on the back at the age of 1 month. Moreover, these same animals were easily approached in the paddock and stroke on the whole body by a familiar or non familiar experimenter one year later while it was even difficult to merely approach the controls at that time (Henry *et al.*, 2003, 2005).

Conclusion

New research lines are arising in the field of human - foal relation and the results obtain up to now show that many elements, in the daily routines of breeding farms may interfere. It is still very difficult to apprehend what is really positive from the animal's point of view. A balance has obviously to be found between too much and too little handling, enough social contact without preventing human contact. It is obviously easy to make mistakes while intentions were positive.

The results obtained in the studies on imprinting indicate that the stress induced is not justified by long lasting positive effects. The use of specific stimuli does not help to prevent the foals to be afraid of them later. On the contrary, weaning and the following year may well be privileged time for establishing the bond.

Finally, the best approach may well be that relying on the most natural and simple bases. Thus, whereas in traditional practices, foals were taken with dams while they worked, the importance of the dam as a social model is mentioned in current practices (unpubl. data).

In natural conditions, the dam seeks isolation from the social group before parturition and prevents her conspecifies to approach the foal for several hours at least after birth, a time period that may be therefore important for the establishment of the mare - foal bond, and thus probably not the best for a human interference (Hausberger *et al.*, 2006.). Moreover it is the first social model that can help acquiring knowledge on the environment. Rather than an early interference, the use of this natural tendency of the foal to learn from the dam seems therefore promising, especially as it does not induce any stress. The durability of the effects obtained after very short handlings of the foal.

It is probable, however, that the effects may very negative if the human - dam relation is aversive! Establishing a positive human - dam relation may therefore be one key to enhance the manageability of foals.

References

Ainsworth, M.D.S., M.C. Blehar, E. Waters and S. Wall, 1978. Patterns of attachment: a psychological study of the strange situation. Hillsdale, Erlbaum.

Belyaev, D.K. and L.N. Trut, 1975. Some genetic and endocrine effects of selection for domestication in silver foxes. In: The wild canids, M.W. Fox (Ed.). Van Nostrand Reinhold, New York, 416-426.

Bertin, A. and M.A. Richard-Yris, 2004. Mothers' fear of human affects the emotional reactivity of young in domestic Japanese quail. Appl. Anim. Behav. Sci., 89, 215-231.

Boissy, A. and M.F. Bouissou, 1988. Effects of early handling on heifer's subsequent reactivity to humans and to unfamiliar situations. Appl. Anim. Behav. Sci. 20: 259-273.

Christensen, J.W., J. Ladewig, E. Sondergaard and J. Malmkvist, 2002. Effects of individual versus group stabling on social behaviour in domestic stallions. Appl. Anim. Behav. Sci. 75: 233-248.

Dawkins, M.S., 1980. Animal suffering: the science of animal welfare, London: Chapman and Hall.

Fordyce, G., M.E. Goddard, R. Tyler, G. Williams and M.A. Toleman, 1985. Cited by Le Neindre *et al.*, 1996.

Hausberger, M. and C. Muller, 2002. A brief note on some possible factors involved in the reactions of horses to humans. Appl. Anim. Behav. Sci. 76: 339-344.
Hausberger, M. and M.A. Richard-Yris, 2003. Approches humaines de la relation Homme / animal: l'exemple de la communication Homme / cheval. In: Homme / Animal: Quelles relations ? Quelles communications ? Publications des Universités de Rouen et du Havre, 71-82.
Hausberger, M. and M.A. Richard-Yris, 2005. Individual differences in the domestic horse, origins, development and stability. In: *The domestic horse,* D. Mills and S. Mc Donnell (Eds). Cambridge: Cambridge University Press, 33-52.
Hausberger, M., C. Bruderer, N. Le Scolan and J.S. Pierre, 2004. Interplay between environmental and genetic factors in temperament/personality traits in horses (*Equus caballus*). J. Comp. Psychol., 118, ss presse.
Hausberger, M., N. Le Scolan, C. Muller, E. Gautier and A. Wolff, 1996. Caractéristiques individuelles dans le comportement du cheval: prédictibilité, facteurs endogènes et environnementaux. In: 22e Journée d'Etude de la Recherche Equine. CEREOPA, Institut du cheval, Paris: 113-123.
Hausberger, M., S. Henry, C. Larose, M-A. Richard-Yris, 2006. First suckling: a crucial event for mother-young attachment? J. Comp. Psychol., in press.
Hediger H., 1968. The Psychology And Behaviour Of Animals in Zoos And Circuses. Dover Publications, Inc, USA.
Heird, J.C., D.D. Whithaker, R.W. Bell, C.B. Ramsey and C.E. Lokey, 1986. The effects of handling at different stages on the subsequent learning ability of yearling horses. Appl. Anim. Behav. Sci. 15: 15-25.
Heleski, C.R., A.C. Shelle, B.D. Nielsen and A.J. Zanella, 2002. Influence of housing on weanling horse behavior and subsequent welfare. Appl.Anim.Behav.Sci. 78: 291-302.
Hemsworth, P.H. and J.L. Barnett, 1992. The effects of early contact with humans on the subsequent level of fear of human in pigs. Appl. Anim. Behav. Sci. 35: 83-90.
Hemsworth, P.H., J.L. Barnett and G. J. Coleman. 1992. Fear of humans and its consequences for the domestic pig. In: The Inevitable Bond: Examining Scientist-Animal Interactions, H. Davis and D. Balfour (Eds.), Cambridge University Press, Cambridge, U.K, 264-284.
Henry, S., M. Hausberger and M.A. Richard, 2003. La mère comme médiateur de la relation à l'Homme: une étude expérimentale chez le cheval. In: 29e Journée d'Etude de la Recherche Equine. Institut du cheval, 26 février, pp 61-70.
Henry, S., D. Hemery, M.-A. Richard and M. Hausberger, 2005. Human-mare relationships and behaviour of foals toward humans. App. Anim. Behav. Sci. 93(3-4): 341-362.
Jezierski, T., Z. Jaworski and A.Gorecka, 1999. Effects of handling on behaviour and heart rate in Konik horses: comparison of stable and forest reared youngstock. Appl. Anim. Behav. Sci. 62: 1-11.
Jones, R.B., 1994. Regular handling and the domestic chick's fear of human beings: generalisation of response. Applied Animal Behaviour Science 42, 129-143.
Jones, R.B. and J.M. Faure, 1981. Tonic immobility ('righting time') in laying hens housed in cages and pens. Applied Animal Ethology 7, 369-372.
Karsh, E.B. and D.C. Turner, 1988. The human - cat relationship. In: The domestic cat: the biology of its behaviour, D.C. Turner and P. Bateson (editors), Cambridge University Press, Cambridge, 159-177.
Kersten, A.M.P., F.M. Meijsser and J.H.M. Metz, 1989. Effects of early handling on later open-field behaviour in rabbits. Applied Animal Behaviour Science 24, 157-167.
Lansade, L., M. Bertrand, X. Boivin and M.F. Bouissou, 2004. Effects of handling at weaning on manageability and reactivity of foals. Applied Animal Behaviour Science 87, 131-149.
Le Neindre, P., G. Trillat, J.M. Chupin, P. Poindron, A. Boissy, P. Orgeur, X. Boivin, J.N. Bonnet, J. Bouix and B. Bibe, 1994. Les ruminants et l'homme, un vieux lien qu'il faut entretenir. Un point sur comportement et bien-être animal. Ed. INRA, Paris.
Lorenz, K., 1935. Der kumpan in der umwelt des vogels, die artgenosse als auslösendes moment sozialer verhaltung weissen. J. Ornithol. 83: 137-213.
Lyons, D.M., E.O. Price and G.P. Moberg, 1988. Social modulation of pituitary-adrenal responsiveness and individual differences in behavior of young domestic goats. Physiol. Behav. 43: 451-458.
Magnusson, D. and R.B. Cairns, 1995. Developmental science: an integrated framework. In: Developmental science, R.B. Cairns, G.H. Elder Jr. and E.J. Costello (editors), Cambridge University Press, New York.

Mal, M.E., C.A. McCall, K.A. Cummins and M.C. Newland, 1994. Influence of prewearing handling methods on post-wearing learning ability and manageability of focals. Appl. Anim. Behav. Sci. 40: 187-195.

Mal, M.E. and C.A McCall, 1996. The influence of handling during different ages on a halter training test in foals. Appl. Anim. Behav. Sci. 50: 115-120.

McGreevy, P.D., N.P. French and C.J. Nicol, 1995a. The prevalence of abnormal behaviours in dressage, eventing and endurance horses in relation to stabling. Veterinary Record 137: 36-37.

Mc Greevy, P.D., P.J. Cripps, N.P. French, L.E. Green and C.J. Nicol, 1995b. Management factors associated with stereotypic and redirected behaviour in the Thoroughbred horse. Equine Veterinary Journal 27: 86-91.

Miller, R.M., 1991. Imprint training of the newborn foal. Western Horseman Inc., Colorado Springs, CO, 44-87.

Murphy, L.B. and I.J.H. Duncan, 1978. Attempts to modify the responses of domestic fowl toward human beings. II - The effect of early experience. Applied Animal Ethology 4: 5-12.

Nicol, C.J., 1995. The social transmission of information and behaviour. Appl. Anim. Behav. Sci. 44, 79-98.

Rivera, E., S. Benjamen, B. Nielsen, J. Shelle and A.J. Zanella, 2002. Behavioural and physiological responses of horses to initial training: the comparison between pastured versus stalled horses. Appl. Anim. Behav. Sci. 78: 235-252.

Ruiz-Miranda, C.R. and M. Callard, 1992. Effects of the presence of the mother on responses of domestic goat kids (*Capra hircus*) to novel inanimate objects and humans. Appl. Anim. Behav. Sci. 33: 277-285.

Sato, S., H. Shiki and F. Yamasaki, 1984. The effects of early caressing on later tractability of calves. Japanese Journal of Zootechnical Science, 55, 332-338.

Scott, J.P. and J.L. Fuller, 1965. Genetics and social behaviour of the dog. University of Chicago Press, Chicago.

Seligman, M.E., S.F. Maier and J.H. Geer, 1968. Alleviation of learned helplessness in the dog. Journal of Abnormal Psychology, 73(3), 256-262.

Simpson, B.S., 2002. Neonatal foal handling. Appl. Anim. Behav. Sci. 78: 303-317.

Sondergaard, E. and U. Halekoh, 2003. Young horses' reactions to human in relation to handling and social environnement. Appl. Anim. Behav. Sci. 84: 265-280.

Sondergaard, E. and H. Schougaard, 2000. The effect of social environment on feed intake, growth and health in young Danish Warmblood horses. In: Book of Abstracts of the 51st Annual Meeting of the European Association for Animal Production, The Hague, the Netherlands, 21-24 August.

Waring, G.H., 1983. Horse behaviour. The behaviour traits and adaptations of domestic and wild horses, including ponies. Noyes publ., New Jersey.

Williams, J.L., T.H. Friend, M.J. Toscano, M.N. Collins, A. Sisto-Burt and C.H. Nevill, 2002. The effects of early training sessions on the reactions of foals at 1, 2 and 3 months of age. Appl. Anim. Behav. Sci. 77: 105-114.

Wright, J.C., 1983. The effects of differential rearing on exploratory behavior in puppies. Applied Animal Ethology 10, 27-34.

Effects of period, type and duration of handling on manageability, reactivity and learning ability of horses

L. Lansade[1], X. Boivin[2] and M.F.Bouissou[1]
[1]INRA, Centre de Recherche de Tours/Nouzilly,37380 Nouzilly, France
[2]INRA, Centre de Recherche de Clermont-Ferrand/Theix, 63122 Saint Genès Champanelle, France

Abstract

Manageability, reactivity and learning ability of horses are three essential characteristics for the rider-horse relationship. In comparison to routine handling, extra handling may reduce fear of humans and general reactivity and thereby improve manageability and learning ability in a variety of farm animal species, including horses. The purpose of this paper is to review the experiments which have studied the effects of handling on manageability, reactivity and learning ability of foals. Many authors have studied the period during which the handling is performed, and particularly the periods following birth and weaning. Even if it has become fashionable to handle foals immediately after birth, experiments which tested the effects of neonatal handling found either short-term effects or no effect. On the other hand, the time of weaning appears to be a favourable period to handle foals, since handling performed at this time provides long term effects on manageability and reactivity. The type of handling and how it is perceived by the animal could also influence the effectiveness of the handling treatment. Many studies have investigated this aspect on other farm animal species. For example, 'forced human handling' of sheep has been compared to 'gentle handling', and only the gentle handling was found to affect approachability. However, to our knowledge, no study on horses has examined the effect of the type of handling and how it is perceived by the animal. Finally, it is also important to consider the duration of the handling treatment and the necessity to maintain its effect over time with additional handling. Until now it has not been proved that handling horses for less than 5 days is effective, on the contrary, two weeks is sufficient to induce some effects and is generally the most common duration used. Handling procedures that use the optimal type, length and period may have durable effects on manageability, reactivity and learning ability of foals. Nevertheless, other approaches such as indirectly approaching the foal through the mother may also have beneficial effects on horse behaviour.

Keywords: horse, handling, manageability, reactivity, human-animal relationship

Introduction

Manageability, reactivity and learning ability of horses are three essential characteristics for the rider-horse relationship. We define manageability as the ease with which a person can impose a routine handling procedure on horses such as fitting a halter, picking up feet or leading. Various authors working with farm species (e.g. cattle) prefer the term tractability, ease of handling or docility (Le Neindre *et al.*, 1996; Simpson, 2002; Lansade *et al.*, 2004). Fearfulness, a major sub-category of emotional reactivity, is defined as a characteristic of the individual animal that leads it to react with the same trend to a variety of frightening events. As demonstrated in several farm species, a high fear level may increase the risk of accidents for handlers and animals, and may affect various aspects of behaviour and productivity (for review: Boissy, 1998; Hemsworth, 2003; Lensink *et al.*, 2000). In addition, high levels of fearfulness impair learning ability (Fiske and Potter, 1979) and therefore may affect the horse's utilisation for riding purposes since it requires that the animal learns many tasks.

Early experience is well known to strongly influence adult behaviour. In comparison to routine handling, extra handling may reduce fear of humans and general reactivity and thereby improve manageability and learning ability in a variety of animal species including horses, although results are sometimes contradictory. For example, some authors have reported lower reactivity for handled animals (rat: Schaeffer, 1963; rabbit: Kersten *et al.*, 1989; silver fox: Pedersen and Jepessen, 1990), whereas others have reported no difference between handled and non-handled subjects (rat: Galef, 1970; hens: Jones and Faure, 1981). The differences observed may be explained by the variety of protocols used. Particularly, the period during which handling is performed, the type and the length of handling are three interlinked factors which can influence strongly the effects of handling. The aim of this paper is to review the experiments which have studied the effects of different handling procedures on manageability, reactivity and learning ability of foals and to discuss the importance of the period, the type and the length of handling. From a practical point of view, the general purpose of the different experiments is to determine which procedures are the most efficient and the most feasible for horse breeders or owners. From a theoretical point of view, these experiments attempt to elucidate the mechanisms underlying the observed effects of handling.

Period of handling

Many authors have focused their studies on the timing of handling, since this may be a particularly important factor in determining whether or not such handling will influence behaviour. The authors have generally considered two periods: the neonatal period and the period immediately following weaning.

Neonatal period

Some authors have suggested the possible existence of a 'sensitive' or 'critical' period in mammals during which the animal is more receptive to external stimuli and to handling. Most studies have favoured the neonatal period as the ideal time for handling (for review: Boivin *et al.*, 2001; Krohn *et al.*, 2001). For instance, Scott (1992) reported that dogs should be handled from 3 to 12 weeks of life in order to be socialised to human. In rats, handling during the first week of life is more effective than at any other period of life (Schaeffer, 1963). In pigs, handling from birth to 3 weeks of age is more effective than from 3 to 6 weeks, or from 6 to 9 months of age (Hemsworth and Barnett, 1992). Young farm ruminants (sheep, cattle, goat) seem more affected in the first days of life by extra human contact than later (Markowitz *et al.*, 1998; Krohn *et al.*, 2001).

A training method known as 'imprint training' (Miller, 1991) has been widely promoted and used in the horse industry. Supporters of this method assert that neonatal handling has very long term effects, lasting for several years, and suggest an imprinting-like process, defined by Lorenz (1957) as an innate behaviour pattern involving an irreversible process of rapid attachment to specific objects during a critical period. Some studies have examined the efficacy of this neonatal handling procedure in horses, although their results are inconclusive. Three of the five studies which compared a neonatal handled group with a non-handled group found that neonatal handling is totally inefficient. Williams *et al.* (2002, 2003) handled foals (rubbing, picking up feet, haltering …) at birth, or / and at 12, 24, 48 hours after birth. They concluded this early training was ineffective since there is no difference between handled foals and a group of non-handled foals when tested at 1, 2 and 3 months of age (Williams *et al.*, 2002) or at 6 months of age (Williams *et al.*, 2003). These results are in accordance with those of Mal *et al.* (1994), which showed no significant difference between foals handled twice daily from birth to 7 days, foals handled twice daily from birth to 7 days plus once weekly until weaning and foals that received no handling, when given a learning task or when rated on their manageability at approximately 4 months of age.

However, other studies found some effects of neonatal handling. Simpson (2002) showed foals handled intensively from 2-8 hours after birth and continuing daily for 5 days were more tractable and less reactive than untreated controls when tested at 4 months of age. However, only a short term effect of neonatal handling has been demonstrated in this study. Lansade *et al.* (2005) tested the long term effects of neonatal handling (over a 1-year period) on manageability, reactivity and learning ability. In accordance with Simpson (2002), they showed that foals handled from 6 hours of age daily for 14 days were more tractable during tests performed at 4 months of age than non-handled foals. However, the effects of neonatal handling decrease rapidly: six months later only few differences were found between handled and non-handled foals and one year later there was no difference between groups. In addition, there did not show effects of handling on reactivity or learning ability.

To summarise, scientific studies to date have demonstrated either no significant effect of neonatal handling (Williams *et al.,* 2002, 2003; Mal *et al*., 1994) or only short term effects (Simpson, 2002; Lansade *et al*., 2005). These differences may be explained by the fact that the length of handling varied considerably between studies: from only one session (Williams *et al*., 2002, 2003) to 40 days (Mal and McCall, 1996). This aspect will be discussed below. However, none of the above studies found evidence of an imprinting-like process, as proposed by Miller (1991). Williams *et al.* (2003) suggested that it is very likely that the early training sessions practised by some horse breeders or owners appear to be effective because they are reinforced frequently by subsequent training sessions.

Even if neonatal handling provided only short term effects, some studies show that they were more effective than later handling. For example, Mal and McCall (1996) found that handling throughout the first 42 days of life increased foals' performance on a halter-training task compared to handling from 43 to 84 days of age. The authors suggested that the importance of timing in handling may be due to the possible existence of a critical period during which handling improves later tractability, or to a phenomenon similar to learned helplessness (to be discussed below). Early handled foals could be restrained more completely during handling due to their smaller size, and may have learned that fighting against restraint and pressure resulted in no change in their situation, and therefore learned non-responsiveness. Because late-handled foals were larger when handled, they could struggle more effectively against restraint and may have learned to fight pressure to a greater degree. However, early handled foals did learn to submit to halter pressure during the training sessions, and consequently it is unlikely they were exhibiting learned helplessness. On the other hand, because early handled foals were more readily restrained during handling, they may have learned that a more pleasant situation followed cessation of struggling and transferred that learning to the halter training task. The larger late-handled foals took more effort to restrain and may not have learned that a more pleasant situation followed when they stopped struggling. A third hypothesis concerns the fact that very young foals have less avoidance reactions towards humans than older foals, certainly due to a maturational process. That may explain why handling after birth is easier and more efficient than later handling.

To conclude this part, even if effects of neonatal handling seem to be only temporary, handling foals around birth may be advantageous because it is more efficient to handle them during this period, and it is also easier to do so due to their smaller size. Nevertheless, some authors (e.g. Simpson, 2002) discuss the potential risks of handling neonatal horses: the dam-young relationship could be broken, the handler may be subjected to aggression from a protective mother, and later in life the animals might not 'respect' humans and may consider them as conspecifics (play, dominance behaviour).

Weaning period

Other developmental periods, such as the time of weaning should therefore also be considered. Artificial weaning (abrupt and complete separation between dam and young) is known to induce stress in many species, including horses (McCall *et al*., 1985; Malinowski *et al*., 1990; Houpt *et al*., 1984). Bateson (1979) suggests that any period of reorganisation, associated with stress, could be a period of special sensitivity to external stimuli. In cattle (Boivin *et al.*, 1992) and goats (Boivin and Braastad, 1996), human contact during this period results in more efficient and friendly human-animal relationships. Lansade *et al.* (2004) compared two groups of foals handled daily for 12 days: the first group was handled immediately after weaning (early handled), whereas handling of the second group began 3 weeks after weaning (late handled). A third group consisted of non-handled foals. Handling consisted of haltering, gently patting all parts of the body, picking up feet and leading. Results showed that it was easier to handle foals immediately after weaning than 3 weeks later. During manageability tests conducted 2 days, 4, 7 and 10 months after the end of the handling period, early and late handled foals were easier to handle and less reactive than controls. During manageability tests performed 18 months after the end of the handling period, the late handled foals did not differ from controls, whereas the early handled foals did. In addition, early handled foals had a lower increase in cardiac rate during a surprise test than late handled foals or controls during tests performed 2 days, 7 and 10 months after the handling period (this test was not conducted 18 months after handling). According to the authors, the period immediately following weaning is therefore a particularly favourable period for handling.

Several hypotheses may be offered to explain why weaning is a favourable period for the effects of handling. The first hypothesis concerns the social reorganisation induced by weaning. In cattle, Veissier and Le Neindre (1989) established that weaning strengthened social relationships between heifers. Moreover, Boivin and Braastad (1996) suggested that the weaning period was favourable for establishing a positive relationship between goat kids and humans because 'at this period kids could be more susceptible to form new social bonds, motivated either by the need for peers or the need for being adopted by a new mother'. They proposed that 'the human might serve also as a surrogate mother'. In horses, a similar process might exist. However, in the study by Boivin and Braastad (1996), kids were weaned precociously at one week of age, whereas the foals in this study were weaned at six months of age, when they were already relatively independent from their mothers. Thus the hypothesis put forward for kids cannot totally explain the results of this experiment. Moreover, it is possible that handling of earlier weaned foals would be even more effective. The second hypothesis concerns learning ability, which is reported in cattle to be better just after weaning than one month later (Veissier *et al.*, 1989). To our knowledge, no such study has been conducted with horses. However, if the same phenomenon exists, it could partly explain the higher efficacy of handling at this period. A final hypothesis is that the stress induced by weaning (Malinowski *et al.*, 1990) is limited to one or two weeks (Houpt *et al.*, 1984; McCall *et al.*, 1985). Thus only early handled foals would have been under stress during handling. Bateson (1979) suggests that any period of reorganisation, associated with stress, could be a period of special sensitivity to external stimuli. Furthermore, this stress might have induced a state of 'learned helplessness' as defined below. The trauma in this situation was induced by the separation from the mother, and the foal learns that resistance is ineffective in reducing stress factors. The foal no longer reacts to its environment and thus would become easier to handle.

To conclude, the different experiments tend to demonstrate that handling around birth has no effect or only temporary effects. On the contrary, effects of handling after weaning appear to be more persistent. Although direct comparisons between the different studies are difficult because of breed differences and differences in environmental conditions, neonatal handling nevertheless seems to be less effective. The presence of the mother could be responsible for this effect in the case of newborn foals. It has been demonstrated in sheep that the presence of the mother may be a factor that limits the establishment of human-animal relationships (Boivin *et al.*, 2001). It would be interesting to compare accurately the effects of handling during the neonatal period with those performed just after weaning.

As we have discussed above, the periods which are the most often studied are those following a natural event (birth) or an artificial event (weaning). Other favourable periods for handling following an artificial event may exist. For example, during the breaking in procedure, some breeders isolate the animal from the rest of the herd to provoke a social rupture, since they empirically found that this rupture may facilitate the handling procedure. The mechanisms on which this phenomenon is based may be similar to those proposed to explain why the period immediately following weaning is a favourable period for handling. In the case of weaning, the rupture concerns the mare-foal bond, whereas in the present case, the rupture concerns the bonds with conspecifics. It would be interesting to test the effects of such social rupture on the effectiveness of handling procedures. More generally, any period of stress (changing place, changing social partners, etc.) may correspond to favourable periods for handling and may be interesting to study.

As mentioned above, the differences observed between results obtained in the different studies (presence or absence of effects) may be explained by the variety of protocols used. In particular, the type and the length of handling may affect the efficiency of the treatment. These two factors are discussed below.

Type of handling

The type of handling and how it is perceived by the animal could strongly influence its effect on the later human-animal relationship. A common hypothesis is that if handling is perceived as an aversive experience by the animal, the animal quickly learns to avoid areas and people associated with such experiences. No study has tested this hypothesis in horses. However, in domestic ewes, Mateo *et al.* (1991) compared the effects of three differential handling procedures on behaviour. The first group was exposed to 'gentle handling', the second group to 'forced human handling' and the third was a non-handled group. Handling treatment was performed for three weeks. Animals underwent a test of approachability to the handler, movement through a chute, halter-restraint in proximity to the handler and simulated shearing. They found that only gentle handling improved approachability during the test. Forced human handling, commonly assumed to be an aversive treatment, produced no effect.

In most previously described experiments performed on horses, handling has consisted in a forced human contact: foals were caught and restrained and the different stimuli (rubbing, picking up feet, presentation of plastic bag, etc.) were imposed on the animal without the possibility of avoidance. This procedure could be assumed to be aversive treatment. However, many studies show that it is effective. Consequently, other mechanisms may explain why this type of procedure can be effective.

Firstly, as Mal and McCall (1996) suggested, the existence of a phenomenon of learned helplessness may explain the efficacy of this procedure: the animal may learn that fighting against restraint and pressure results in no change in its situation, and therefore it learns non-responsiveness. According to Overmier and Seligman (1967), 'learned helplessness might well result from receiving aversive stimuli in a situation in which all instrumental responses or attempts to respond occur in the presence of the aversive stimuli and are of no avail in eliminating or reducing the severity of the trauma'. A lack of response-reinforcement contingency during training may result in non-responsiveness in a separate, though similar, situation.

Another hypothesis is that the foal rapidly learns that the different stimuli do not constitute a threat, and progressively ceases to react to them. For instance, farm animals tend to habituate to repeated management procedures and thus learn to accept the restraint involved with such handling, no longer reacting when they are confronted with them again. Grandin (1989) allowed sheep that had previously experienced both electro-immobilisation and restraint in a squeeze tilt table to approach the table to be restrained again. After a time, the sheep voluntarily accepted the tilt table restraint, with some even lining up to re-enter the system.

It would be of interest to test different types of handling horses to identify which is the most adapted to them. In addition, the age of the animal may influence which type of handling is the most suitable. For example, as Mal and McCall (1996) suggested, the phenomenon of learned helplessness may be greater in young foals (0-42 days of age) than older animals (43-82 days of age) because it is more difficult to restrain older foals. Moreover, testing which type of handling is the most effective requires understanding how it is perceived by the animal. For example, the respective role of feeding or gentling to improve the human - farm animal relationship is still under discussion even if feeding per se is a strong reward in a conditioning process (for review, Boivin *et al.*, 2003). It could be of interest to study handling preferences in a Y-maze as Pajor *et al.* (2003)'s study of dairy cattle, in which animals can choose between various handling treatments (shouting *versus* gentle speaking, hand feeding *versus* pail feeding etc.). In the above experiment, it has been shown that shouting and the use of cattle prods are aversive, whereas tail twisting is not if done gently; feeding is rewarding although, for heifers, feeding by hand may not be. Finally, they found no evidence that gentling or being spoken to in a gentle voice is rewarding for cattle. Useful information could be obtained by carrying out a similar study with horses.

Length of handling

The last factor examined in this paper is the length of handling. The optimal length of handling must be both long enough to be effective and short enough to be feasible for horse breeders. In horses, few studies to date have attempted to determine the optimal length of handling. The optimal length of handling also depends on the period and the type of handling. A short period of handling performed at the optimal time, using the most appropriate method, may be more efficient than a longer period at a less favourable time or using an inappropriate method. Authors have handled foals for lengths of time between one day and several months, and it can be observed that the degree of efficiency of the different handling procedures studied is generally dependent on the length of the treatment.

All the studies in which foals are handled for several months show that the handling is usually effective. For example, Heird *et al.* (1981) reported that foals which had frequent human contact before weaning and were handled immediately after weaning once a day for a week and then handled weekly, or foals handled four times a week for ten months, scored higher on learning tests at 14 months of age than non-handled foals. Jezierski *et al.* (1999) reported that intensively handled foals (10min, 5 days per week, beginning from the age of 2 weeks or ten months to the age of 24 months) scored better on manageability tests (catching, leading, picking up feet, being approached by an unfamiliar person) and demonstrated lower heart rates than those not handled. Other studies show that the longer the handling period, the more efficient the procedure is. Heird *et al.* (1986) compared 5 groups: limited handling, 1 week of handling, 2 weeks of handling, 3 weeks of handling, continuous handling for 18 months. At two years old, the horses were tested in a simple place-learning T-maze. The most-handled group reached a consistently higher percentage of correct responses earlier than the less-handled groups. It was also the least emotional, as determined by the animals' reactivity to a novel stimulus, and received the highest scores for trainability after being ridden. Therefore these different studies show that handling animals for several months can improve learning ability and manageability. However, many breeders do not have the time to handle their foals for such a length of time. Consequently, other authors have handled foals during shorter periods from a few weeks to a few days, and have found some effects. McCann *et al.* (1988) based the length of the handling period on the work of Heird *et al.* (1986), which demonstrated that the learning curves from the T-maze performance of yearlings reached asymptote by Day 10. Therefore, they chose to handle foals daily for 14 days. Foals were handled daily in 20-min sessions (haltered, taught to lead, secured to a post, rubbed over the body and head) and compared with non-handled foals. They found that handled yearlings tended to exhibit lower heart rates with the handling stimulus than non-handled yearlings. All the procedures for handling foals for this two-week period were effective, as those used by Lansade *et al.* (2005) and Lansade *et al.* (2004) where animals were handled for a two-week period, either just after birth or just after weaning. On the other hand, handling for less than one week is not usually enough to be effective. However, Simpson (2002), who handled animals daily for the first five days of life, found that it was sufficient to provide some effects, whereas Mal *et al.* (1994) found that daily handling for 7 days just after birth was not effective, and Williams *et al.* (2003, 2004) also found that handling animals daily for the first four days of life was ineffective.

To conclude, several months of handling are effective but this is not feasible for horse breeders, whereas handling for less than one week is not usually sufficient. Thus, a two-week handling period seems to be a good compromise between feasibility and efficacy. However, it is important to underline again the fact that the optimal length depends on the period and type of handling.

Conclusion

This paper examines three important and interlinked factors which can determine the efficacy of a handling procedure: the period, the type and the length. The different studies that have focused on the effects of the period suggest to date that neonatal handling has no effect or only short term effects. These results do not corroborate the hypothesis of an imprinting phenomenon in horses claimed by Miller, who has widely promoted the neonatal handling method in the horse industry. In fact a more favourable time to handle foals seems to be the period following weaning: it is easier to handle them during this period than later and the effects persist at least 18 months after the end of the handling. Other periods such as those immediately following a stressful event (e.g. social rupture) may also be favourable times for handling and merit further investigation. Several hypotheses have been proposed to explain why some periods may be more favourable for handling than others: existence of a critical period, the learned helplessness phenomenon, the maturational process etc., but none have so far been tested in horses. In addition, to our knowledge, no study has been performed to determine the most effective type of handling for horses, also the perception the animal has of it is particularly important for the efficacy of the procedure. Finally, with regard to the length of handling, this probably depends on the period of life and the type of human contact. A two-week period of handling could be a good compromise between feasibility and efficacy, but this needs more detailed investigation. In addition to the handling treatment, other approaches such as environmental enrichment (corresponding to the improvement in the biological functioning of captive animals resulting from modifications to their environment, Newberry, 1995) may also have beneficial effects on horse behaviour and require further investigation. Indirect approaches of the foal through the mother could also be particularly beneficial: Henry *et al.* (2005) showed that handled mares during the first five days of their foals' lives facilitates human-foal relationship until at least one year later. In this case, a such treatment appears even to provide more durable effects than directly handle the foal.

References

Bateson, P., 1979. How do sensitive periods arise and what are they for? Animal Behaviour, 27: 470-486.

Boissy, A., 1998. Fear and fearfulness in determining behavior. In: Genetics and the behavior of domestic animals. T. Grandin (editor),Academic Press San Diego 67-111.

Boivin, X. and B.O. Braastad, 1996. Effects of handling during temporary isolation after early weaning on goat kids' later response to humans. Appl. Anim. Behav. Sci., 48: 61-71.

Boivin, X. J. Lensink, C. Tallet and I. Veissier, 2003. Stockmanship and farm animal welfare. Animal Welfare, 12: 479-492.

Boivin, X., R. Nowak and A.T. Garcia, 2001. The presence of the dam affects the efficiency of gentling and feeding on the early establishment of the stockperson-lamb relationship. Appl. Anim. Behav. Sci., 72: 89-103.

Boivin, X., P. Le Neindre, J.M., Chupin, J.P. Garel and G. Trillat, 1992. Influence of breed and early management on ease of handling and open-field behaviour of cattle. Appl. Anim. Behav. Sci., 32: 313-323.

Fiske, J.C. and G.D. Potter, 1979. Discrimination reversal learning in yearling horses. J. Anim. Sci., 49: 583.

Galef Jr., B.G., 1970. Aggression and timidity: responses to novelty in feral norway rats. J. Comp. Physiol. Psychol. 70: 370-381.

Grandin, T., 1989. Voluntary acceptance of restraint by sheep. Appl. Anim. Behav. Sci., 23: 257-261.

Heird, J.C., A.M. Lennon and R. W. Bell, 1981. Effects of early experience on the learning ability of yearling horses. J. Anim. Sci., 53: 1204-1209.

Heird, J.C., D.D. Whitaker, R.W. Bell, C.B. Ramsey and C.E. Lokey, 1986. The effect of handling at different ages on the subsequent learning ability of 2-year-old horses. Appl. Anim. Behav. Sci., 15: 15-25.

Hemsworth, P.H., 2003. Human-animal interactions in livestock production. Appl. Anim. Behav. Sci., 81: 185-198.

Hemsworth, P.H. and J.L. Barnett, 1992. The effects on early contact with humans on the subsequent level of fear of human in pigs. Appl. Anim. Behav. Sci., 35: 83-90.

Henry, S., D. Hemery, M.-A. Richard and M. Hausberger, 2005. Human-mare relationships and behaviour of foals toward humans. App. Anim. Behav. Sci. 93(3-4): 341-362.

Houpt, K.A., H.F. Hintz and W.R. Butler, 1984. A preliminary study of two methods of weaning foals. Appl. Anim. Behav. Sci., 12: 177-181.

Jezierski, T., Z. Jaworski and A. Gorecka, 1999. Effects of handling on behaviour and heart rate in Konik horses: comparison of stable and forest reared youngstock. Appl. Anim. Behav. Sci., 62: 1-11.

Jones, R.B. and J.M. Faure, 1981. The effects of regular handling on fear in the domestic chick. Behav. Process. 6: 135-143.

Kersten, A.M.P., F.M. Meijsser and J.H.M. Metz, 1989. Effects of early handling on later open-field behaviour in Rabbits. Appl. Anim. Behav. Sci. 24: 157-167.

Krohn, C.C., J.G. Jago and X. Boivin, 2001. The effect of early handling on the socialisation of young calves to humans. App. Anim. Behav. Sci., 74: 121-133.

Lansade, L., M. Bertrand, X. Boivin and M.-F. Bouissou, 2004. Effects of handling at weaning on manageability and reactivity of foals. App. Anim. Behav. Sci., 87: 131-149.

Lansade, L., M. Bertrand and M.-F. Bouissou, 2005. Effects of neonatal handling on subsequent manageability, reactivity and learning ability of foals. App. Anim. Behav. Sci., 92: 143-158

Le Neindre, P., X. Boivin and A. Boissy, 1996. Handling of extensively kept animals. App. Anim. Behav. Sci., 49: 73-81.

Lensink, B. J., X. Boivin, P. Pradel, P. Le Neindre and I. Veissier, 2000. Reducing veal calve's reactivity to people by providing additional human contact. J. Anim. Sci., 78: 1213-1218.

Lorenz, K., 1957. Companionship in bird life; fellow members of the species as releasers of social behavior. In: Instinctive behavior. C. H. Schiller (C. H. Schillers),University Press New York.

Mal, M.E. and C.A. McCall, 1996. The influence of handling during different ages on a halter training test in foals. Appl. Anim. Behav. Sci., 50: 115-120.

Mal, M.E., C.A. McCall, K.A. Cummins and M.C. Newland, 1994. Influence of preweaning handling methods on post weaning learning ability and manageability of foals. Appl. Anim. Behav. Sci., 40: 187-195.

Malinowski, K., N.A. Hallquist, L. Helyar, A.R. Sherman and C.G. Scanes, 1990. Effect of different separation protocols between mares and foals on plasma cortisol and cell-mediated immune response. J. Eq. Vet. Sci., 10: 363-368.

Markowitz, T.M., M.R. Dally, K. Gursky and E.O. Price, 1998. Early handling increases lamb affinity for humans. Anim. Behav., 55: 573-587.

Matéo, J.M., D.Q. Estep and J.S. McCann, 1991. Effects of differential handling on the behaviour of domestic ewes. App. Anim. Behav. Sci., 32: 45-54.

McCall, C.A., G.D. Potter and J.L. Kreider, 1985. Locomotor, vocal and other behavioral responses to varying methods of weaning foals. Appl. Anim. Behav. Sci., 14: 27-35.

McCann, J.S., J.C. Heird, R.W. Bell and L.O. Lutherer, 1988. Normal or more highly horses. II. The effects of handling and reserpine on the cardiac response to stimuli. Appl. Anim. Behav. Sci., 19: 215-226.

Miller, R.M., 1991. Imprint training of the newborn foal. Western Horseman Inc.Colorado Springs Co.

Newberry, R.C., 1995. Environmental enrichment: Increasing the biological relevance of captive environments. Appl. Anim. Behav. Sci., 44: 229-243.

Overmier, J.B. and M.E.P. Seligman, 1967. Effects of inescapable shock upon subsequent escape and avoidance responding. J. Comp. Physio. Psycho., 63: 28-33.

Pajor, E.A., J. Rushen and A.M.B. de Passille, 2003. Dairy cattle's choice of handling treatments in a Y-maze. Appl. Anim. Behav. Sci., 80: 93-107.

Pedersen, V. and L.L. Jeppessen, 1990. Effects of early handling on later behaviour and stress responses in the Silver Fox (Vulpes vulpes). Appl. Anim. Behav. Sci. 26: 383-393.

Schaeffer, T., 1963. Early experience and its effects on later behavioral processes in rats: II. A critical factor in the early handling phenomenon. In: Critical periods. J. P. Scott (editors), Dowden, Hutchinson and Ross, Inc Stroudsburg, Pennsylvania 272-290.

Scott, J.P., 1992. The phenomenon of attachment in human-non human relationships. In: The inevitable bond; examining scientist-animal interactions. H. David, Balfour, D. (editors),Cambridge University Press Cambridge 72-92.

Simpson, B.S., 2002. Neonatal foal handling. Appl. Anim. Behav. Sci., 78: 303-317.

Veissier, I. and P. Le Neindre, 1989. Weaning in calves: its effects on social organization. Appl. Anim. Behav. Sci., 24: 43-54.

Veissier, I., P. Le Neindre and G. Trillat, 1989. Adaptability of calves during weaning. Biology of Behaviour, 14: 66-87.

Williams, J.L., T.H. Friend, M.J. Toscano, M.N. Collins, A. Sisto-Burt and C.H. Nevill, 2002. The effects of early training sessions on the reactions of foals at 1, 2 and 3 months of age., Appl. Anim. Behav. Sci., 77: 105-114.

Williams, J.L., T.H. Friend, M.N. Collins, M.J. Toscano, A. Sisto-Burt and C.H. Nevill, 2003. Effects of imprint training procedure at birth on the reactions of foals at age six months. Equine Vet. J., 35: 127-132.

Investigations on the activity of trotters and the relationship to growth

L. Voswinkel[1], K.-H. Tölle[2], C. Henze[1], K. Blobel[3] and J. Krieter[1]
[1]Institute of Animal Breeding and Husbandry, Christian-Albrechts-University, D-24098 Kiel, Germany
[2]Chamber of Agriculture Schleswig Holstein, D-24327 Blekendorf, Germany
[3]HIPPO-Blobel, D-23812 Ahrensburg, Germany

Abstract

The aim of this study was to determine factors affecting the activity of juvenile horses on pasture during rearing. Furthermore the influence of the activity on growth parameters and on basis endurance was analysed. Data from 30 horses -16 mares and 14 stallions- of one stud were available. After weaning, two foals were first housed in single boxes (pair) with daily exercise for 4 to 5 weeks. Afterwards all weanlings were separated by sex and kept in group housing with direct access to pasture. Pedometers and videotaping were used to measure the activity. Climatic data were taken each test day. To determine individual development, growth parameters were recorded. To analyse the endurance, heart rate measurements and lactate values were used. During the investigation period the individual activity of each horse remained constant compared to the average of the herd. Pairs of foals showed a more similar activity pattern ($p<0.001$). Positive phenotypic correlation was found between activity and diameter of superficial digital flexortendon ($r_p=0.24$) and deep digital flexortendon ($r_p=0.27$). A negative correlation was estimated between the activity of the animals and their lactate value ($r_{animal} = -0.40$). Foals with low activity during rearing tended towards higher lactate values at the beginning of the training.

Keywords: horse, activity, growth, endurance

Introduction

Endurance is the basis to practice any other physical abilities (Lindner, 1997). Therefore training of endurance is important to avoid damages and injuries caused by overwork, irrespective of the discipline (Schäfer, 2000).

The organism of the horse is strongly affected by specialisation towards a run- and escape animal. Under close-to-nature conditions, horses move about 16 hours per day (Kolter and Meyer, 1986; Kiley-Worthington, 1990).

The objective of the study was to examine which factors have an impact on the activity of juvenescent horses. The influence of the activity during rearing on growth and on performance was also to be determined. It was hypothesised that a greater activity affects the development of bones and tendons positively and results in a higher basic endurance.

Material and methods

Data of 30 trotters, born between 08.02.2004 and 06.06.2004, were available. Data recording occurred in the period from November 2004 to June 2006. At the beginning of the training, data acquisition (April, 2006) 18 animals were available. To ensure that all animals had grown up under nearly identical environmental conditions, the investigation was carried out on one stud. After weaning two foals were first housed together in single boxes with daily exercise for 4 to 5 weeks. Afterwards, the weanlings were seperated by sex (16 females, 14 males) and kept together in two groups in loose housing systems with direct access to pasture. After an adaptation-phase of four weeks, measurements of activity started by using pedometers. The measurand of a pedometer is the number of actions of the foreleg of a horse. According to this, the below-mentioned activity per hour is the quotient of the totally measured actions of the foreleg and the time in hours between two metre- reading dates. To analyse different movement types, digital cameras recorded the action of the foals on every investigation day (n = 112 investigation days). For intervals of ten weeks, the development in height at withers, cannon bone diameter and circumference, length of foreleg, weight, and diameter of flexortendons were recorded. To examine the individual bone development, X-rays of the epiphysis of the distal radius were taken at the time of weaning and at the beginning of the training in May, 2006. The different stages of epiphyseal closure were divided into 5 classes (Table 1).

During the training unities (8km per unity) the heart rate, the speed and the distance were recorded by a combined heart rate/Global Positioning System (GPS) (Garmin GPS Forerunner 301). Five minutes after the load, a blood sample was taken from vena jugularis externa. The lactate value was determined by a quick test (Accurend® Lactate). Climatic data of each test day were available from a weather station.

Table 2 shows an overview about averages, standard deviation as well as the minimum and the maximum of chosen parameters.

The statistical analysis was performed with the software package SAS (SAS, 2005). The influence on the activity was estimated by using a mixed linear model with the fixed effects temperature (≤ 2.5 °C, > 2.5 °C to ≤ 5 °C, > 5 °C), rain (0 ml, > 0 ml), wind speed (< 3 m/s, ≥ 3 m/s) and their interactions. The activity was the repeated performance of each animal. For this reason, the animal as well as the test day were taken into consideration as random effects. On account of the weaning

Table 1. Classes of the grade of epiphyseal closure.

Class	Explanation
I	No periostal reaction in the distal radius - epiphyseal plate completely open
II	Low periostal reaction/structural compressions in the distal radius - epiphyseal plate completely open
III	Zones of compression in the distal radius - recognisable narrowing of the epiphyseal plate
IV	High reactions in the distal radius/high bone compression - recognisable beginning of epiphyseal closure
V	Completed epiphyseal closure

Table 2. Chosen parameters with their average (Mean), standard deviation (Std), minimum (Min) and maximum (Max).

Parameter	Mean	Std	Min	Max
Activity per hour (act/h)	65.2	16.6	36.5	129.9
Superficial digital flexortendon (cm)	2.0	0.2	1.7	2.4
Deep digital flexortendon (cm)	1.4	0.2	1.0	1.9
Grade of epiphyseal closure	3.0	1.1	2.0	5.0
Pace (km/h)	24.0	6.7	2.0	45.0
Heart rate (bpm)	108.0	36.0	35.0	239.0
Lactate value (mmol/l)	2.2	0.6	1.4	3.7

method with two foals per single box, the random effect of the pair also was considered. To estimate the influence on growth parameters, the fixed effect sex and the random effects animal and test day were included in model. On each day of growth measurement, the horses were at different ages, so the age within the test day was regarded as covariable. For the determination of the influence on performance, endurance parameters were corrected by the fixed effects birth season (early = February/March, late = April/May) sex, distance (km) and the random effect animal. Phenotypic correlations between activity and growth were estimated by residuals. The relationship between activity and endurance parameters was determined by animal effects.

Results and discussion

The investigation showed a significant influence of the temperature and the wind speed on the activity. Higher temperature increased activity, while a stronger wind speed decreased activity (Table 3). For the interpretation of these results it has to be considered, that most of the data were collected during the winter period and therefore very high temperatures did not appear.

In spite of nearly identical environmental conditions the horses showed differences in height of activity (Table 2). The repeatability was very high (74.9%) and marked a constant movement pattern of the individual animal. The random effect weaning pair had a high influence on activity, so the height of activity adapted in each case to both animals of a pair.

The physiological age of epiphyseal closure in the distal radius is 24 to 30 months (Huskamp *et al.*, 1996). At the beginning of training, the horses were at an age of 22 to 27 months. The evaluation of the radiographs showed a distribution of the horses on the different grades of epiphyseal closure. 17% of the horses were in class III, 39% in class IV and 44% of the animals had completed bone growth in the distal radius (class V). Estimates of activity in the different stages of growth plate closure showed that horses with lower activity during rearing were more often in class III at the beginning of training and horses with higher activities tended towards a completed epiphyseal closure in the distal radius (Table 4).

Table 3. Least-Square-Means (LSM) and their standard errors(SE) and the significance level(Pr > F) of the parameter activity per hour (act/h) for the effects temperature and wind speed.

Parameter	Class	LSM (act/h)	SE	Pr > F
Temperature (°C)	≤2.5	56.67	3.92	<0.0001
	>2.5 to ≤5	59.24	3.96	
	>5	71.80	3.47	
Wind speed (m/s)	<3	64.13	3.33	<0.0001
	≥3	61.01	3.74	

Table 4. Mean estimates of activity (act/h) in the different stages of epiphyseal closure.

Class		Act/h
III	Epiphyseal plate open	54.5
IV	Beginning of epiphyseal closure	57.6
V	Total epiphyseal closure	70.8

In an investigation in forty-three foals Barneveld and van Weeren (1999) found that a lack of exercise generally delayed the developement of the musculoskeletal system.

In addition to the relationship between activity and bone growth the influence on tendon growth was analysed. Positive phenotypic correlations were found between activity and diameter of superficial digital flexortendons (r_p = 0.27) as well as the diameter of deep digital flexortendons (r_p = 0.24). In comparison, Cherdchutham *et al.* (1999) showed that 5 month-old foals kept at pasture had significantly stronger tendons structures than foals kept in box rest and box rest with enforced exercise. In another study, Kasashima *et al.* (2002) used an increasing amount of treadmill exercise for a small periode each day in addition to 4 hours of pasture exercise in foals from 6 weeks to 15 month of age beside a control group, which received 4 h pasture exercise. A significant increase in the rate of growth in the cross-sectional area of the superficial digital flexortendon with treadmill exercise was documented. These results show that early activity of the growing horse is an important determinant of tendon development.

Hardly any relationship could be determined between the activity during rearing and the heart rate at the beginning of training (r_{animal} = 0.09). A negative correlation was estimated between activity and lactate value (r_{animal} = -0.40), so that animals with low activities during rearing tended to have higher lactate values at the beginning of training.

Arnemann (2003) found significantly higher heart rates and lactate values in boxed horses than in loose-housed horses. This supports the hypothesis that a higher free activity leads to better endurance values.

Conclusion

The distribution of the classes of epiphyseal closure and the positive coherence between activity and the diameter of superficial and deep digital flexortendons show the dependence of osseous and tendon development from biomechanical effects, which has to be considered in the start date of training. An expected low heart rate of horses with high activities during rearing could not be confirmed. The strong dependence of heart rate to mental stress was possibly the reason for this result. More resistant to psychical factors is the lactate value. Negative relationships between activity and lactate value showed that horses with higher activity start in the training with better basic endurance. These first results show the importance of activity during rearing as a factor which enables horses to become most resistant to injury during adulthood.

References

Arnemann, S., 2003. Haltung von Sportpferden unter besonderer Berücksichtigung der Leistung. Dissertation Hannover.

Barneveld, A. and P.R. van Weeren, 1999. Conclusions regarding the influence of exercise on the development of the equine musculoskeletal system with special reference to osteochondrosis. Equine Veterinary Journal Supplement. 31:112-9.

Cherdchutham, W., C. Becker, R.K.W. Smith, A. Barneveld and P.R. van Weeren, 1999. Age-related changes and effect of exercise on the molecular composition of immature equine superficial digital flexor tendons. Equine Veterinary Journal Supplement. 31:86-94.

Huskamp, B., K. Dämmrich, J. Erbslöh and L.B. Jeffcott, 1996. Skelettreife und Trainingsbeginn bei Vollblutpferden unter besonderer Berücksichtigung des Tierschutzgesetzes. Hrsg.: Direktorium für Vollblutzucht und Rennen wak Verlag und Kunstberatung, München.

Kasashima, Y., R.K.W. Smith, H.L. Birch, T. Takahashi, K. Kusano and A.E. Goodship, 2002. Exercise-induced tendon hypertrophy: cross-sectional area changes during growth are influenced by exercise. Equine Veterinary Journal Supplement. 34:264-268.

Kiley-Worthington, M., 1990. The behavior of horses in relation to management and training - towards ethologically sound environments. Journal of Equine Veterinary Science. 10: 1. 62-71.

Kolter, L. and H. Meyer, 1986. In: Unterlagensammlung Pferdehaltung; 1. Ernährung und Haltung. Deutsche Reiterliche Vereinigung (Hrsg.); FN Verlag, Warendorf.

Lindner, A., 1997. Laktat und Leistung; Verlag Dr. A. Lindner.

SAS, 2005. SAS/STAT Users Guide, Version 8.2.

Schäfer, B., 2000. Reaktionen physiologischer Leistungskriterien auf zusätzliches Ausdauertraining während der reiterlichen Ausbildung von Sportpferden. Dissertation Göttingen, FN-Verlag, Warendorf.

Part C

Importance of social environment on the behaviour of young and adult horses

The effects on behaviour of the presence or absence of a familiar horse during feeding

S.E. Redgate[1] and H.P.B. Davidson[2]
[1]IERM, University of Edinburgh West Mains Road, EH9 3JG United Kingdom, Current address: Animal Behaviour Cognition and Welfare Group. Department of Biological Sciences, University of Lincoln, Riseholme Hall, LN2 2LG, United Kingdom; sredgate@lincoln.ac.uk
[2]Equine Studies Group, c/o WALTHAM Centre for Pet Nutrition, Leicestershire LE14 4RT United Kingdom.

Abstract

Increasing visual contact between stables improves awareness of conspecifics, which has been linked to a decrease in abnormal behaviour. There has however, been limited investigation into the relationship between neighbouring stabled horses especially during periods of excitement such as meal times. Seven adult riding horses of mixed breed were used; six acted as observers and one as a demonstrator. Three treatments were applied: (1) (stable): observer eating alone; (2) (arena): observer eating, demonstrator not eating; (3) (arena): observer and demonstrator eating. Horses were fed a standardised plain concentrate feed; behaviour was recorded using instantaneous sampling every 5 seconds for the 5-minute test period. Individual differences existed in the amount of vigilant behaviour recorded; three horses' showed more vigilant behaviour in the presence of another horse either eating or not eating than when stabled alone. Conversely, the second three horses' showed more vigilant behaviour when stabled alone. In 4 of the 6 horses' studied as vigilant behaviour increased time taken to consume the feed decreased. These trends suggest that sub groups may exist in the population. The results have implications for stable design. For certain horses during competitive activities such as feeding, the close presence of another horse may be either beneficial or potentially stressful. As this is only a preliminary investigation, further research is advocated.

Keywords: horse, stable, neighbour, feeding

Introduction

From domestication to the present day the horse's role has changed substantially. During the last 100 years alone this has varied from the working horses in agriculture and transport to the leisure and sports horses of today. However, the practice of housing each horse individually in an enclosed stable has changed little. Physical contact between horses is avoided for many reasons including: to minimise the risk of injury, prevention of disease outbreak, to ease handling and to have control over individual food intake.

As a gregarious species, the horse's survival has been dependent on the cohesion and effectiveness of its social structures. However, the horse is often housed individually and therefore, deprived of social contact. Even when visual access was allowed between neighbouring stables, through the provision of grills, when turned out to pasture horses did not associate more with their former neighbour (Christensen *et al.* 2002). For the horse the stable is therefore, a particularly restrictive social environment. McGreevy *et al.* (1995a) found that time spent in the stable was correlated with an increased risk of abnormal behaviour. However, if visual contact was allowed between stable interiors, through the use of bars and grilles, there was a reduced risk of abnormal behaviour developing (McGreevy *et al.* 1995b). Furthermore, increasing the visual access to the surrounding environment and conspecifics or simply providing a mirror or a poster to mimic the presence of a conspecific has reduced the incidence of certain stereotypies, specifically weaving (Cooper *et al.* 2000, McAfee *et al.* 2002 and Mills and Riezebos, 2005). When grazing as a group, horses tend to associate with preferred group members and distance themselves from others (McGreevy, 2004). However, when stabled a horse's nearest neighbour is often not accounted for when deciding on housing arrangements. Cooper *et al.* (2000) noted an increase in arousal during meal times. How this affects behaviour of neighbouring horses in stabling which has visual access has not been established.

The practice of opening up stables to improve visual access to conspecifics has been taken as beneficial to equine well being. There has, however, been limited investigation into the relationship between neighbouring stabled horses especially during potentially exciting periods such as meal times. The objective of this study was to investigate the effects on behaviour of the presence or absence of a familiar horse during meal times.

Materials and methods

Three treatments' were tested, on three separate occasions one week apart at the same time each day. Throughout the trial the procedure was standardised and each horse acted as its own control. The horses were given one session per week for three weeks' prior to the data collection days to adjust to the trial equipment, the presence of another horse and eating in the trial arena.

Experimental subjects

Seven riding horses of mixed breed were used from the Easter Bush Riding Stables, University of Edinburgh. Six of the horses were geldings and one a mare, aged between 8 - 21. All horses were familiar with each other and had been resident at the yard for at least 9 months prior to the start of the study. Each horse was stabled in a traditional enclosed loose box; conspecifics could be viewed from the door. Horses were stabled for 24hrs except for daily ridden exercise and free turnout to grass once a week.

One horse acted as a demonstrator and six acted as observers. The demonstrator was chosen due to the horse's quiet nature and willingness to eat.

Experimental arena and equipment

Three treatments' were tested, as shown in Table 1. Six brick loose-boxes of a similar design and familiar to each horse were used for treatment 1. Each box only had a view from the door to other stabled companions and the courtyard. An indoor arena that all horses were familiar with was used for treatments 2 and 3.

Two adjacent pens (4 x 7m) were constructed in the arena from electric fencing, without any charge running through. This allowed full visual, olfactory and auditory contact between the horses. A feed bucket was placed at the centre front of each pen. The bucket was weighted with a tyre so that any spilt food was minimised. A video camera mounted on a tri-pod was positioned so that both pens could be viewed and filmed equally. Treatment 1 was not filmed and observations were recorded straight onto a check sheet.

Table 1. Description of the three treatments used in the trial.

Treatment	Description
1	Observer alone in its own stable
2	In the arena, observer horse without the demonstrator horse eating
3	In the arena, observer horse with the demonstrator horse eating

A plain high fibre coarse mix was used for the trial. The horses had prior experience of this feed in their daily feeding regimen. A standard amount of 200g was weighed out for each horse.

Experimental protocol

Each horse was tested according to a latin square and observed for 5 minutes starting from the moment the horse was released. Focal animal sampling was used with instantaneous sampling every 5 seconds for the duration of the 5-minute period. The ethogram used to record the behaviour of both the observer and demonstrator horses is shown in Table 2. During test sessions the same researcher was consistently present but remained as discreet as possible. The feed was prepared before the horses entered the arena to minimise any cues and the pens were checked for debris left by the previous horse. The bucket was positioned at the front of the pen before each horse was released from a pre-determined point. A cleaned empty bucket and tyre was positioned at the front of the pen during treatment 2. Horses had not received any food for at least 1-hour prior to the start of the trials. In order to motivate the demonstrator to eat during each test session, three horses were tested in treatment 2 and three in treatment 3 during the first week; this was alternated for the second week.

Table 2. Ethogram of the behaviour patterns observed during the trial.

Behaviour	Description
Eat	Head in bucket chewing
Eat hover (semi-vigilant)	Neck below the withers, head hovers over the bucket whilst chewing
Eat head up (vigilant)	Neck above the withers, head up whilst chewing
Investigate	Sniff or lick the tyre or feed bucket

Data analysis

Observer reliability over time was analysed from three randomly selected observation periods (Martin and Bateson, 1993). Data was checked for normality and analysed using Minitab version 12.1. Differences between the three treatments were assessed using a One-Way ANOVA. Where the normal distribution was not met a Kruskal Wallis test was applied.

Results

Observer reliability between the three observation periods proved to be consistent; the percentage agreement was 92%, 97% and 95%. No significant differences existed for each of the behaviours across the three treatments' for the trial (Table 3). However, there were individual differences for the behaviours of eat hover and eat head up.

Table 3. Summary information for each behaviour pattern. Values listed are mean ± S.E.M.

Behaviour	Treatment 1	Treatment 2	Treatment 3
Eat	15.17±3.40	18.83±3.06	21.17±2.24
Eat head up	4.67±2.23	5.00±1.29	5.00±1.15
Eat hover	7.67±2.39	8.17±1.66	5.33±2.09
Investigate	6.67±2.63	4.67±2.06	3.17±1.11

Individual differences

The behaviours of eat hover and eat head up were both considered as vigilant behaviour however, eat head up was taken to provide a clear indication of vigilance. Half of the group (n=3) showed a decrease in frequency of the behaviour head up when observed in the stable compared to the arena. However, the other half of the group increased the frequency of head up in the stable compared to the arena. In 4 out of the 6 horses' studied an increase in vigilant behaviour was accompanied by a decrease in the time taken to consume the food (range between 16 and 71 seconds).

Discussion

Vigilance is behaviour that increases the probability that an animal will detect a given stimulus at a given time for example, to detect the presence of a predator or a dominant group member (Dimond and Lazarus, 1974). Eat head up and eat hover were interpreted as vigilant behaviour as the horse would raise the head and neck whilst eating in order to survey the environment.

The horses in this study were split into two groups, depending on the amount of vigilant behaviour shown in the two different test situations of the stable and the arena. In the stable visual access to conspecifics was restricted in comparison to the arena where another horse was always within close proximity.

Observations showed that the close presence of a group member during feeding decreased vigilant behaviour in half of the horses studied. This is in agreement with the work by Sweeting *et al.* (1985) they also found that when stabled horses were given visual access to a conspecific in the next stable the incidence of vigilance reduced. Studies with a variety of species have reported that the frequency of vigilant behaviour decreases when group size increases (Roberts, 1996). This may be due to an increase in predator detection as there are more individuals to look out for predators. Sullivan (1984) reported that the time woodpeckers spent 'looking out' decreased and time spent eating increased when foraging in a mixed species flock. When the head down field of view was obstructed in feeding birds head up vigilance increased (Guillemain *et al.* 2001). These horses increased their vigilance when stabled in order to watch any activity from the stable door.

The other half of the group of horses studied showed an increase in vigilance when a group member was present. This group of horses decreased their vigilance when stabled individually. It may be that these horses perceived the demonstrator to be a threat and therefore increased their vigilance. Unfortunately, the social hierarchy was not known for this group of horses. One of the strongest influences on group feeding behaviour is the horse's rank within the group (McGreevy, 2004). Kolter (1984) demonstrated that during feeding of group housed horses the hierarchy between the horses whilst feeding was based on the degree of tolerance from the dominant members of the group. Additionally, the construction and close position of the pens in the arena could have had an influence on behaviour. Holmes *et al.* (1987) demonstrated that the design and use of a partition between subordinate and dominant animals when feeding in close proximity affected time spent feeding.

Regardless of the presence of another horse in the arena the test environment could have been regarded as novel, the horses were regularly exercised in this area, not fed. Furthermore, feeding tends to be a competitive activity especially when a high value feed such as a concentrate mix is offered. Therefore, both the environment and feed may have heightened the need to remain vigilant for certain horses.

Ultimately, individual differences in behaviour did not significantly have an effect on the amount of feed consumed and hence probably would not affect production measures such as growth rate. However, all six horses' showed vigilant behaviour either in the arena or the stable, four of these horses showed a decrease in the time taken to consume the food. A decrease in time to eat the food may mean that the horse was eating the food too quickly and may increase the risk of getting choke.

The results of this study indicate that the environment and the presence of a familiar horse can affect vigilance whilst feeding. Further work is needed to investigate if this effect persists over a longer period of time with a typical concentrate meal (1000g).

The social hierarchy was not established for this particular group of horses further studies could manipulate the status of neighbouring horses in order to evaluate vigilant behaviour further.

The differences recorded in vigilance highlight the importance of treating the individual when housing horses. Social contact for the horse is a necessary component of normal horse behaviour; therefore it should be incorporated into stable design. However, for certain horses during excitable periods such as feeding the close proximity of another horse may be a potential stressor. Investigating the behaviour of neighbouring horses through various stable designs that allow visual and social contact is advocated in order to cater for the individual's social and behavioural needs.

Acknowledgements

This work was completed as part of an MSc at the University of Edinburgh. The author would like to thank the Equine Studies Group, WALTHAM Centre for Pet Nutrition for their support, Dr. Natalie Waran for supervision during the project and Jo Hockenhull and Rhoda Sylvester-Evans for technical assistance.

References

Christensen, J.W., J. Ladewig, E. Søndergaard and J. Malmkvist, 2002. Effects of individual versus group stabling on social behaviour in domestic stallions. Appl. Anim. Behav. Sci. 75: 233-248.

Cooper, J.J., L. MacDonald and D.S. Mills, 2000. The effect of increasing visual horizons on stereotypic weaving: implications for the social housing of stabled horses. Appl. Anim. Behav. Sci. 69: 67-83.

Dimond, S. and J. Lazarus, 1974. The problem of vigilance in animal life. Brain Behav. Evol. 9: 60-79.

Guillemain, M., P. Duncan and H. Fritz, 2001. Switching to a feeding method that obstructs vision increases head-up vigilance in dabbling ducks. J. Avian Biol. 32: 345-350.

Holmes, L.N., G.K. Song and E.O. Price, 1987. Head partitions facilitate feeding by subordinate horses in the presence of dominant pen-mates. Appl. Anim. Behav. Sci. 19: 179-182.

Kolter, L., 1984. Social relationships between horses and its influence on feeding activity in loose housing. Proceedings of the International Congress of Applied Ethology in Farm Animals. Darmstadt, Kiel, 151-155 pp.

Martin, P. and P. Bateson, 1993. Measuring Behaviour. An introductory guide. Second Edition, Cambridge University Press, 120 pp.

McAfee, L.M., D.S. Mills and J.J. Coooper, 2002. The use of mirrors for the control of stereotypic weaving behaviour in the stabled horse. Appl. Anim. Behav. Sci. 78: 159-173.

McGreevy, P.D., N.P. French and C.J. Nicol, 1995a. The prevalence of abnormal behaviours in dressage, eventing and endurance horses in relation to stabling. Vet. Rec. 137: 36-37.

McGreevy, P.D., P.J. Cripps, N.P. French, L.E. Green and C.J. Nicol, 1995b. Management factors associated with stereotypic and redirected behaviour in the Thoroughbred horse. Equine Vet. J. 27: 86-91.

McGreevy, P.D., 2004. Equine Behavior. A Guide for Veterinarians and Equine Scientists. Saunders, 194pp.

Mills, D.S. and M. Riezebos, 2005. The role of the image of a conspecific in the regulation of stereotypic head movements in the horse. Appl. Anim. Behav. Sci. 91: 155-165.

Roberts, G., 1996. Why individual vigilance declines as group size increases. Anim. Behav. 51: 1077-1086.

Sullivan, K.A., 1984. The advantages of social foraging in downy woodpeckers. Anim.Behav. 32: 16-24.

Sweeting, M.P., C.E. Houpt and K.A. Houpt, 1985. Social facilitation of feeding and time budgets in stabled ponies. J. Anim. Sci. 60: 369-374.

The effects of social environment on the behaviour of young horses

E. Søndergaard and J.W. Christensen
Danish Institute of Agricultural Sciences, Research Centre Foulum, 8830 Tjele, Denmark

Abstract

This paper is an overview of the first large study aimed at investigating specific effects of different social environments on the development of young horses. Forty Danish Warmblood colts were used in two replicates of 20 horses from weaning until 2½ years of age. In each replicate, 8 horses were housed singly, and 12 horses were housed in 4 groups of 3 horses during the housing period from September to May. The horses had daily access to paddocks, either singly (singly stabled) or in groups (group stabled). The summer period was spent on pasture in groups. Singly housed horses showed more interest in contacting humans and were more easily approached by humans in their home environment. Singly housed horses interacted more with a trainer during weekly training sessions than group housed horses. The interaction consisted mainly of non-aggressive biting, indicating that singly housed horses were motivated for physical interaction. However, singly housed horses completed fewer stages in the training program and they also bit and kicked more during training sessions than group housed horses. Group housed horses exercised more in their paddocks compared to singly housed horses, and it is therefore very likely that group housed horses develop a better co-ordination of movements than singly housed horses. Likewise, observations on social behaviour during the summer season showed that previously group stabled stallions had a higher frequency of displacements and submissive behaviours but fewer direct aggressive interactions, compared to previously singly housed stallions, implying that group housed stallions had a more refined social language. Although horses are social by nature their social skills still have to be refined and practised. The differences between the treatment groups imply that by housing young horses singly, thus not giving them the opportunity to practice their social skills they may be more prone to injuries when interacting with other horses as adults. The study concludes that group housing is essential for young horses in relation to their social behaviour and also beneficial in relation to their reactions in training situations.

Keywords: young horses, social behaviour, housing, training

Introduction

The behavioural development of a horse from birth till adulthood is a continuous process that needs the presence of certain environmental features, as well as the absence of others, to occur optimally. For some behaviour patterns (e.g. maternal bonding) exposure must occur during specific periods in order to exert an effect, but for other behaviour patterns (e.g. socialisation on conspecifics or humans) repeated exposure over an extended period appears be to necessary in order for learning to be permanent (Ladewig *et al.*, 2005). Since these ontogenetic mechanisms seem to operate similarly under natural as well as husbandry conditions, optimal housing and management of the young developing horse is necessary in order to obtain an adult horse with desirable reactions.

The social environment of young domestic horses is often a matter of being raised alone or with young horses of the same age, which is far from the natural life of young horses where they live in family groups until the age of 2 to 3 years and then in either bachelor groups or mixed sex juvenile groups (Waring, 2003). The social environment constitutes an important part of the housing environment of young horses but other parts of the housing such as spacing (e.g. Zeitler-Feicht and Prantner, 2000) or access to exercise (e.g. Bell *et al.*, 2001) have attracted more focus in scientific studies. However, in a Danish study the focus was to investigate the effects of the social environment and handling on the behavioural and physical development of young horses from weaning until 2½ years of age. To our knowledge, this is the largest project so far on the specific effects of different social environments on the development of young horses.

The present paper will refer mainly to this experiment, but also to other relevant studies.

Description of experiment

The experiment was conducted at the Danish Institute of Agricultural Sciences, Research Centre Foulum, Denmark. Forty Danish Warmblood colts were used in two replicates of 20 horses. Twenty Danish Warmblood (DW) male foals were purchased from 19 private farms in 1997 and 1999, and brought to the experimental station at weaning at an age of 4.3 (± 0.5) and 5.0 (± 0.5) months of age, respectively. All foals were born in April or May and raised mainly on pasture with their dams and no access to additional feed. The foals were offspring from 6 sires, 4 sires in each replicate. From the time of purchasing at an age of approximately 2 months until weaning the foals were handled minimally. At the age of 3.0 ± 0.5 months all foals were tested in a reactivity test as described by Søndergaard (1998).

At Research Centre Foulum all foals were housed in the same building during the experimental period of two years. In each replicate, 8 horses were housed singly in boxes of $9m^2$, and 12 horses were housed in 4 groups of 3 horses in boxes of $27m^2$. Separation between boxes allowed the horses to see, hear, smell and touch but not physically interact with neighbouring horses. The front of each box consisted of vertical bars with a distance of approximately 30cm allowing the horses to feed from the floor in front of each box. Horses were fed *ad libitum* every morning with a Total Mixed Ration of chopped grass silage, chopped hay, chopped straw and concentrate, molasses and minerals. The Total Mixed Ration was adjusted during the experiment according to the body condition of the horses by increasing the amount of forage and changing the relation between ingredients. Feeding regime and recordings of growth and physical development are described in Søndergaard (2003).

During the housing period, the horses were given 3 hours of daily exercise in paddocks according to their housing, i.e. single housed horses alone and group housed horses in their groups of 3. The size of the paddocks was 20 x 40 m for singly housed horses and 45 x 90 m for group housed horses. The horses were housed from mid-September to mid-May. The first summer period was spent on pasture in one group whereupon they were again housed singly or in the same groups of three during the following winter period. In the second summer period, singly housed horses were pastured in one group and group housed horses in another group for six weeks due to studies on social behaviour (Christensen *et al.* 2002a, b) whereupon the groups were joined for another 4 weeks. In August after the second summer period the horses were housed and handled until the end of the experiment in the middle of September. During the housing periods several tests on human-animal relationship were performed. These tests are described in Søndergaard and Halekoh (2003).

Half of the singly housed horses and half of the group-housed horses were handled for 10 minutes three times per week during each housing period. Handling involved leading, tying up, touching, lifting legs etc. The handling scheme is described in details in Søndergaard and Ladewig (2004). Non-handled horses were only handled for monthly weightings, farrier and veterinary treatment.

Effects of social environment

General activity and behavioural development

Lack of exercise in young foals has been shown to alter their locomotive behaviour (Barneveld *et al.*, 1999), and retard the development of the musculo-skeletal system by lowering the bone mineral density and alter the chemical composition of tendon and articular cartilage (Barneveld and van Weeren, 1999). Group housed horses exercise more when in a paddock compared to singly housed horses (Søndergaard, 2003). The exercise in group housed horses does not only consist of forward locomotion but also e.g. rearing, bucking and play fighting, and it is therefore very likely that group housed horses develop a better co-ordination of movements than singly housed horses. This may be essential for their later use as sport horses, since Back *et al.* (1995) showed that the kinetics of locomotion observed in foals were predictive of their locomotion as adult horses. However, the results of the Danish study are not in accordance with Kusunose *et al.* (1986), who found that horses that were alone on pasture travelled longer than horses pastured in groups. In their study, however, the horses pastured alone were without horses in neighbouring paddocks and the increased exercise is likely to be a reaction to isolation rather than an effect of the social environment per se. Kusunose *et al.* (1986) also found that group size influenced the activities of the horses. Grazing time and duration of grazing bouts were shorter when only one horse was pastured and increased linearly as the group size increased up to four horses and more.

Even though the horses in the Danish study were only in paddocks for approx. 3 hours per day there was a significant difference in the behaviour shown by the horses, since group housed horses showed a more diverse behaviour repertoire than singly housed horses (Søndergaard, 2003). This is in accordance with Heleski *et al.* (2002) who found that time-budgets of horses pastured in groups of three horses resembled a feral horse time budget whereas singly housed horses spent more time lying and less time moving. Singly housed horses spent significantly more time licking or chewing walls, kicking at the walls, pawing and bucking/rearing bouts (Heleski *et al.*, 2002) indicating that there was a motivation for activity that was not fulfilled. This is confirmed in a study by Hughes *et al.* (2002), who found that solitary foals manipulated objects more than socially raised foals.

When the horses were exposed to a novel environment and isolation at one year of age, group housed horses vocalised more than singly housed horses (Søndergaard and Halekoh, 2003) indicating that they were more distressed. These results are not in accordance with studies on calves where individually housed calves were more fearful when isolated in a novel arena than group housed calves (Jensen *et al.*, 1997). When the test was performed at two years of age there was no difference, possibly due to the horses being more used to be taken away from the other horses e.g. in connection with weightings, veterinary treatments and behaviour tests. Likewise, group housed horses defecated more often when they were alone in the training arena than singly housed horses (Søndergaard and Ladewig, 2004). On the other hand, singly housed horses ran loose for longer periods in the arena than group housed horses before the training sessions started (Søndergaard and Ladewig, 2004). However, the latter result may reflect the fact that the singly housed horses are motivated for exercise or play when in company with humans.

Glade (1984) studied the social sleeping behaviour in young horses and found that when space allowance was not a limiting factor, group housed young horses had fewer but longer periods of recumbency than young horses housed singly. In general, young horses spent more time lying than older animals and for all age groups recumbency was influenced by space allowance (Zeitler-Feicht and Prantner, 2000). Young horses housed singly in stalls spent about one third of the day in a recumbent position. This may be a reflection of a quieter environment in the stalls, or due to the fact that the restriction of the environment limited the possibility to perform other behaviour patterns compared to young horses in outdoor pens, which had the opportunity of interacting with neighbours in the adjoining pens (Glade, 1984). Within groups high-ranking animals tend to lye more than lower-ranking animals (Zeitler-Feicht and Prantner, 2000). In the Danish experiment lying behaviour was recorded in the first replicate for the first two weeks, i.e. right after weaning (Matthiesen, 1999). On the first day after weaning singly housed foals rested in lateral recumbency more often than group housed horses, but by the end of the observation period there was no difference in resting behaviour between singly housed and group housed horses.

Matthiesen (1999) observed feeding activity at treatment level in the first two weeks after weaning, and found that singly housed horses had more eating bouts on day 1, 4 and 7, and ate for longer time on day 4. On day 14 after weaning there was no difference in neither eating frequency nor eating time (Matthiesen, 1999). Kolter (1984) showed that not only rank but also the tolerance of individuals in a group determines the feeding activity. Low-ranking animals get less access to feed, but the tolerance of individuals, which may be different from the rank order, was even more important. In groups, which are not fully integrated, e.g. due to introduction of new animals, some animals may be prevented from feeding to an extent that impairs their welfare (Kolter, 1984). In practice, however, this problem may be solved by head partitions (Holmes *et al.*, 1987) allowing subordinate animals to feed. In an experiment where horses were kept together according to their emotionality, i.e. a 'nervous' group and a 'normal' group, the 'normal' yearlings tended to consume feed in a more regular manner throughout the daylight hours than the 'nervous' yearlings (McCann *et al.*, 1988).

Horse-horse interactions

Although horses are social by nature their social skills still have to be refined and practised. In the Danish experiment the social behaviour of group housed horses seemed more refined than that of singly housed horses as shown by the lower frequency of aggressive behaviour and the higher frequency of subtle agonistic interactions such as displacements and submissive behaviours (Christensen *et al.*, 2002a). These differences imply that by housing young horses singly, thus not giving them the opportunity to practice their social skills they may be more prone to injuries when introduced to other horses. Likewise, singly housed horses performed significantly more play fighting. In a survey on race farms most injuries were attributed to the young horses playing too hard (Gibbs and Cohen, 2001).

Previously group housed colts frequently had a former group mate as their nearest neighbour whereas previously singly housed colts did not associate more with their former box neighbours, when they were pastured in groups (Christensen *et al.*, 2002a). This result indicates that physical contact may be necessary in order to establish bonds between animals, and that full physical contact is an important part of the social behaviour. This is confirmed in calves as measured by operant conditioning, where the calves were more motivated to get access to full contact than to head contact (Holm *et al.*, 2002), indicating that limited social contact is not enough to fulfil the animals' social motivation. Additionally, previously singly housed horses stayed closer together than group housed horses (Christensen *et al.*, 2002a) perhaps indicating a higher level of social motivation, or perhaps due to nervousness, since the same was observed for 'nervous' horses compared to 'normal' horses in the study by McCann *et al.* (1988). Also, Kusunose *et al.* (1986) found that the mean individual distance increased from 5 m when two horses were pastured together up to 30-50 m when 12 horses were pastured together. This may indicate that horses are more comfortable when in a large group but as the mean distance to the nearest neighbour was constant irrespective of group size it may also reflect the fact that when the group gets too large sub-groups will occur.

Singly housed horses spent more time performing social grooming and play behaviour than group housed horses when given the opportunity at pasture (Christensen *et al.*, 2002a) indicating that the motivation to perform these behaviours had been present also during confinement. This was confirmed in the observations of the horses in the paddocks during the housing period, where singly housed horses tried to gain access to social contact over the fence, indicating a strong motivation although they had limited access (Søndergaard, 2003).

Horse-human interactions

In the Danish study, singly housed horses showed more interest in contacting humans and were more easily approached by humans in their home environment (Søndergaard and Halekoh, 2003). This result corresponds with observations by Hughes *et al.* (2002), who found that foals kept without contact to other foals interacted more readily and spent more time with an observer than foals kept on pasture with other foals. Also, in calves, singly housed animals interacted sooner and more often with a test person than calves housed in pairs (Lensink *et al.*, 2001). When the horses were tested for their reactions to an unfamiliar person in an unfamiliar environment no effect of social environment was found (Søndergaard and Halekoh, 2003), which may reflect that the effect of isolation masks other effects. Isolation is a stressful event for a horse, and the presence of a human may not have a calming effect as indicated in a study by Jezierski and Górecka (1999). In their study, horses that were isolated in a stable due to the other horses leaving the stable, had a further rise in heart rate if a human was present, although this may be a conditioning effect, i.e. expectation of being untied and joining the other horses (Jezierski and Górecka, 1999).

In the Danish experiment singly housed horses also interacted more with a trainer during weekly training sessions than group-housed horses (Søndergaard and Ladewig, 2004). The interaction consisted mainly of non-aggressive biting indicating that singly housed horses were motivated for physical interaction. This result was confirmed by observations during summer when horses from both treatment groups were kept on pasture in groups, where singly housed horses spent more time performing social grooming than group housed horses (Christensen *et al.*, 2002a).

In an experiment by Rivera *et al.* (2002) it was found that singly housed horses needed more time for initial training than group housed horses. Since the singly housed horses were kept in stalls, whereas the horses housed in groups were kept on pasture, environmental enrichment and stimulation could be part of the difference in their results. In the Danish experiment, however, similar results were found. Singly housed horses completed fewer stages in the training program and they bit and kicked more during training sessions than group-housed horses (Søndergaard and Ladewig, 2004). There was a large variation between horses in the number of sessions they needed to complete a stage in the training program but this could not be related to the social environment. Whether the performance in the training program was related to tractability or learning ability is unclear. Also, there may be an additional effect of the social environment e.g. hierarchy within the groups although Haag *et al.* (1980) found no correlation between the rank in the hierarchy and performance in two different learning tests in 10 ponies. In a study of calves (Lensink *et al.*, 2001) the contrary result was found, it took more time and effort to load pair housed calves than singly housed calves. The difference may be due to the fact that the calves were loaded individually and were not used to being isolated as opposed to the horses in the Danish study and in the study by Rivera *et al.* (2002), which were always isolated in training sessions.

Conclusion

Group housed horses exercise more than singly housed horses, which is beneficial for the development of the musculo-skeletal system. Group housed horses are less aggressive and their social behaviour seems more refined than that of singly housed horses. Singly housed horses approach sooner and can be approached more closely by humans when tested in their home environment than group housed horses. Group housed horses tended to fulfil more stages in the training program than singly housed horses, and singly housed horses kicked and bit more during training than did group housed horses.

We conclude that group housing has several positive effects on the development and behaviour of young horses.

Perspectives

The Danish study shows that group housing is essential for young horses in relation to their social behaviour and also beneficial in relation to their reactions in training situations. However, it is still an open question whether young horses have to stay in groups all day all year round, or if part-time group housing may have the same positive effects. Another issue is how to achieve the best composition of the group. Most young horses are housed in semi-natural groups, i.e. homogenous groups where a clear leadership may be missing. It is evident that foals learn many things from the dam, e.g. food selection but it is unclear how lack of an older, experienced leader affects the behavioural development of young horses after weaning. Another aspect of the homogenous groups is the fact that the highest-ranking animal may not be psychologically capable of being a 'leader', a fact that may potentially result in a tyrannical leader. This problem has been recognised in practice but remains to be investigated in an experimental study. The use of horses in leisure and sport requires that horses often meet or are housed with unknown horses. How horses are best prepared for such situations is not known. Additionally, adult horses will often be on their own, and therefore it may be useful for young horses to learn to be separated from other horses in various situations.

However, despite the holes in our knowledge there is no doubt from the results of the Danish study and other studies that group housing of young horses and probably of older horses too, should be promoted

References

Back, W., H.C. Schamhardt, W. Hartman, G. Bruin and A. Barneveld, 1995. Predictive value of foal kinematics for the locomotor performance of adult horses. Research in Veterinary Science 59: 64-69.

Barneveld, A. and R. van Weeren, 1999. Conclusions regarding the influence of exercise on the development of the equine musculoskeletal system with special reference to osteochondrosis. Equine vet J., Suppl. 31: 112-119.

Barneveld, A., R. van Weeren and J. Knaap, 1999. Influence of early exercise on the locomotion system. Paper presented at the 50th Annual Meeting of the European Association for Animal Production, Zurich, August 1999

Bell, R.A., B.D. Nielsen, K. Waite, D. Rosenstein and M. Orth, 2001. Daily access to pasture turnout prevents loss of mineral in the third metacarpus of Arabian weanlings. J. Anim. Sci.79: 1142-1150.

Christensen, J.W., J. Ladewig, E. Søndergaard and J. Malmkvist, 2002a. Effects of individual versus group stabling on social behaviour in domestic stallions. Appl. Anim. Behav. Sci. 75: 233-248.

Christensen, J.W., T. Zharkikh, J. Ladewig and N. Yasinetskaya, 2002b. Social behaviour in stallion groups (*Equus przewalskii* and *Equus caballus*) kept under natural and domestic situations. Appl. Anim. Behav. Sci. 76: 11-20.

Gibbs, P.G. and N.D. Cohen, 2001. Early management of race-bred weanlings and yearlings on farms. J. Equine Vet. Sci. 21: 279-283.

Glade, M.J., 1984. Social sleeping behavior in young horses. Equine Practice 6: 10-14.

Haag, E.L., R. Rudman and K.A. Houpt, 1980. Avoidance, maze learning and social dominance in ponies. J. Anim. Sci. 50: 329-335.

Heleski, C.R., A.C. Shelle, B.D. Nielsen and A.J. Zanella, 2002. Influence of housing on weanling horse behavior and subsequent welfare. Appl. Anim. Behav. Sci. 78: 291-302.

Holm, L., M.B. Jensen and L.L. Jeppesen, 2002. Calves' motivation for access to two different types of social contact measured by operant conditioning. Appl. Anim. Behav. Sci. 79: 175-194.

Holmes, L.N., G.K. Song and E.O. Price, 1987. Head partitions facilitate feeding by subordinate horses in the presence of dominant pen-mates. Appl. Anim. Behav. Sci. 19: 179-182.

Hughes, C., D. Goodwin, P.A. Harris and H.P.B. Davidson, 2002. The effect of social environment on the development of object play in domestic horse foals. Proceedings from a Dorothy Russell Havemeyer Foundation Workshop on Horse Behavior and Welfare held at Hólar College - Hólar, Iceland 13-16 June 2002. p.21-22.

Jensen, M.B., K.S. Vestergaard, C.C. Krohn and L. Munksgaard, 1997. Effect of single versus group housing and space allowance on responses of calves during open-field tests. Appl.Anim.Beh.Sci. 54, 109-121.

Jezierski, T. and A. Górecka, A., 1999. Relationship between behavioural reactions and heart rate in horses during transient social isolation. Animal Science Papers and Reports 17: 101-114.
Kolter, L., 1984. Social relationships between horses and its influence on feeding activity in loose housing. Proc. Int. Cong. Appl. Ethol. in Farm Animals. p.151-155.
Kusunose, R., H. Hatakeyama, F. Ichikawa, K. Kubo, A. Kiguchi, Y. Asai, Y. and K. Ito, 1986. Behavioral studies on yearling horses in field environments 2. Effects of the group size on the behavior of horses. Bull. Equine Res. Inst. 23: 1-6.
Ladewig, J., E. Søndergaard and J.W. Christensen, 2005. Ontogeny: preparing the young horse for its adult life. In: The Domestic Horse. The evolution, development and management of its behaviour, D. Mills and S. McDonnell (editors), Cambridge University Press, Cambridge, UK, 139-149.
Lensink, B.J., S. Raussi, X. Boivin, M. Pyykkönen, I. Veissier, 2001. Reactions of calves to handling depend on housing condition and previous experience with humans. Appl. Anim. Behav. Sci. 70: 187-199.
Matthiesen, C., 1999. Comparison of two strategies for weaning of colts. B.Sc. thesis, The Royal Veterinary and Agricultural University, Copenhagen, 1999, 31 p. (In Danish).
McCann, J.S., J.C. Heird, R.W. Bell and L.O. Lutherer, 1988. Normal and more highly reactive horses. I. Heart rate, respiration rate and behavioral observations. Appl. Anim. Behav. Sci. 19: 201-214.
Rivera, E., S. Benjamin, B. Nielsen, J. Shelle and A.J. Zanella, 2002. Behavioral and physiological responses of horses to initial training: the comparison between pastured versus stalled horses. Appl. Anim. Behav. Sci. 78: 235-252.
Søndergaard, E., 1998. Reaction to separation on mares and foals. Proc. of the 32nd International Congress of the ISAE, July 21-25, Clermont-Ferrand, France, 120.
Søndergaard, E., 2003. The effect of social environment and handling on the behavioural and physical development of young horses. Ph. D. thesis. The Royal Veterinary and Agricultural University, Copenhagen, 2003. 123 p.
Søndergaard, E. and U. Halekoh, 2003. Young horses' reactions to humans in relation to handling and social environment. Appl. Anim. Behav. Sci. 84: 265-280.
Søndergaard, E. and J. Ladewig, 2004. Group housing exerts a positive effect on the behaviour of young horses during training. Appl. Anim. Behav. Sci. 87: 105-118.
Waring, G.H., 2003. Horse behavior. Noyes Publications/William Andrew Publishing, Norwich, NY, USA. 442p.
Zeitler-Feicht, M.H. and V. Prantner, 2000. Liegeverhalten von Pferden in Gruppenauslaufhaltung. Arch. Tierz., Dummerstorf 43: 327-335.

Individual paddocks versus social enclosure for horses

G.H.M. Jørgensen and K.E. Bøe
Department of Animal and Aquacultural Sciences, Norwegian University of life Science, P.O. Box 5003, 1432 Ås, Norway; grete.meisfjord@umb.no

Abstract

In order to test the differences in activity and behaviour of horses in either individual or social paddocks, nine (3 cold blood and 6 warm blood) adult horses were turned out in a joint enclosure (2700 m^2), and then the same nine horses were exposed to an individual paddock (10 x 30 m) for 2 hours daily, one week in both treatments. The horses were stabled in tie stalls in the same stable building, and fed 1 hour before turnout. The two last days of the week in the individual paddock and the four last days of the week in the joint enclosure, the horses' behaviour and social activity were observed respectively, using instantaneous sampling every second minute, for two hours. When kept in a social group horses were generally more active, spending less time standing passively (mean ± SE % of total observations; 5.3 ± 2.7% vs. 51.5 ± 6.9%, paired t-test, t= 7.9, P<0.0001), exploring their environment less (0.7 ± 0.4% vs. 3.7 ± 1.1%, t= 2.8, P<0.05) and eating more grass (74.5 ± 3.3% vs. 27.1 ± 8.2%, t= 5.9, P<0.001) than in individual paddocks. Based on the negligible amounts of aggressive interactions and the presence of non- aggressive social behaviours like allogrooming, turnout in social groups appear to be a good alternative to individual paddocks.

Keywords: horse, social, paddock, behaviour, turnout

Introduction

Feral horses in natural or semi-natural environments form family groups with one adult stallion and several mares with their offspring under three years of age (e.g. Klimov, 1988; Boyd, 1991). These family groups' home range are dependent on the resources available, and horses living in dry, mountain areas usually have a larger home range than horses living in lush grasslands (Miller, 1983; King, 2002; King and Gurnell, 2005). Boyd *et al.* (1988) found Przewalski horses travelling for only 7.4% of the 24-hour period when kept in a grassy enclosure.

Regulations for other farm animal species like laying hens, cattle and pigs are moving towards loose housing both in Norway and within the EU. However, the most common way to keep horses is individually in boxes or tied up in stalls (Switzerland: Bachmann and Stauffacher, 2002; Denmark: Søndergaard, *et al.*, 2002; Sweden: Statens Jordbrugsverk, 2004). The main argument for keeping horses in traditional housing systems is the fear of injuries from fights or play in a social group. In spite of this, recent studies show that aggressive episodes and injuries in social herds hardly is any problem (Christensen *et al.*, 2002 b; Grogan and McDonnell, 2005; Lehmann *et al.*, 2006) and that even stallions raised under traditional conditions communicated well with each other and showed social behaviour patterns very similar to behaviours of wild Przewalski stallions when kept together at pasture (Christensen, *et al.*, 2002 a). Group housing of horses involve both lower building costs and larger total space for movement and activity (Ventorp and Michanek, 1997; Søndergaard *et al.*, 2002), and this is also true when using joint enclosures in stead of individual paddocks.

The aim of this experiment was to investigate the effect of individual versus social turnout on behaviour and general activity of domestic horses. We also wanted to look at the number of aggressive interactions between the horses when kept in the social enclosure. It was hypothesised that the activity and the amount of playing and running would increase in a joint enclosure compared to individual paddocks.

Material and methods

Experimental design

A total of nine horses were first grouped together in a joint enclosure for 2 hours daily, in a period of one week. Then the same horses were turned out in an individual paddock for 2 hours a day, for one week. When in the individual paddock, the horse could see and hear other horses, but not have any physical contact with them. The experiment was done in June/July 2004 and the joint enclosure measured 30 x 90 m (2700 m^2, 300 m^2 per horse), whereas the individual paddock measured 10 x 30 m (300 m^2). Both joint enclosure and paddock had a sandy soil ground, with some grass and green plants growing all around the outside perimeter of the fence. In addition some very short grass less than 1 cm high grew on the ground surface inside the joint enclosure.

Animals and feeding

Nine healthy horses (10-22 years old) from both cold- (CB) and warm-blooded (WB) breed types were selected. There were both mares (3) and geldings (6) and all horses were normally used in schooling, either riding or carriage driving. The three CB horses ranged in body weight from approximately 300-500 kg, whereas the six WB horses ranged in body weight from 400-700 kg. The horses were stabled in tie-stalls measuring 1.87 x 2.50 m in the same building, and they were fed individually adjusted rations of good quality hay (6-8 kg per day) four times a day, in addition to oat grain and barley pellets, three times a day and some 3 kg of newly cut forage grass at the evening meal, while housed. In addition the horses got a vitamin and mineral supplement once a day and had free access to salt blocks and water from automatic drinking vessels. The horses were fed approximately 1 hour before turnout to the paddocks. During both experimental treatments the horses did not have any exercise other than their activity in the paddock.

Behavioural observations

When the horses were located in the individual paddock their behaviour and activity was recorded by a present observer, for two hours at the last two days of the week. The following mutually exclusive behavioural classes were scored using instantaneous sampling at two-minute intervals:

- standing (the horse stands inactive with all four feet in contact with the ground or with one foot hooked up in a resting position, relaxed position of the ears and head, but with eyes open);
- walking;
- running (any form of trot, canter or gallop);
- play / play fighting (any form of jumping, buck kicking, rearing or kicking with the feet, not in combination with aggressive display);
- exploration (the horse is standing or walking slowly with the neck horizontally or lower, smelling the ground or an object);
- eating grass;
- standing alert (the horse stands with ears and eyes focused; from Strand *et al.*, 2002);
- self- grooming (an act of integumentary care (e.g., licking, scratching) in this case an animal grooming itself, from Hurnik *et al.*, 1995);
- rolling (an activity characterised by lying down and rolling onto the back accompanied by rubbing movement against the ground, from Hurnik *et al.*, 1995);
- defecation;
- standing asleep (the horse stands in a resting position, on all four legs or with one hindleg hocked, with the head low and eyes closed, dozing);
- digging the ground (pawing, a scooping front leg movement, generally on the ground, from Hurnik *et al.*, 1995);
- eating/ licking soil;

- eating/ licking the wooden fence.

In addition to these behaviours, social behaviours were recorded when the horses were in the joint enclosure, using instantaneous sampling every second minute for two hours, at the four last days of the week.

- aggressive display (flat ears, eye white showing);
- kicks;
- chasing (one or more horses running after and chasing other horses);
- bites;
- greeting behaviour (two horses approach each other, sniffs each other on the muzzle, flanks and/or genitalia, one or both of the horses may vocalise or squeal during this ritual);
- social grooming (allogrooming).

Aggressive display, biting, kicking and chasing was summarised under the term 'aggressive interactions', while play / play fighting, greeting and social grooming was summarised under the term 'non-aggressive social interactions'.

Statistics

From the observations from the two observational days in individual paddock and the four observational days in joint enclosure, mean values for each individual horse was calculated and compared using a paired t-test (two-tailed).

Results

When kept in a social group, the nine horses were generally more active, spending less time standing passively and eating more grass than in individual paddocks (Table 1).

The behaviours vigorous play, digging the ground, rolling, standing asleep, eating/licking fence, eating/licking soil and other behaviours were never observed in the social enclosure and defecation (0.3 ± 0.2%) were rarely observed. Social grooming (allogrooming) was observed 1.7 ± 0.3% of total observations.

Aggressive interactions was rarely observed and non-aggressive interactions tended to be more prevalent after more days spent in the joint enclosure (3.5 ± 0.9%) compared to early in this experimental period (1.5 ± 0.8%, P=0.06) (Figure 1).

Table 1. Activity and behaviour in social enclosure versus individual paddock.

Behaviour (mean ± SE, % of observations)	Joint enclosure	Individual paddock	t- value	P- value
Standing	5.3 ± 2.7	51.5 ± 6.9	7.9	0.0009
Walking	10.3 ± 1.5	9.5 ± 3.8	-0.17	n.s
Exploring	0.7 ± 0.4	3.7 ± 1.1	2.81	0.02
Running	0.8 ± 0.3	0.31 ± 0.26	-1.38	n.s
Eating grass	74.4 ± 3.3	27.1 ± 8.2	5.9	0.0005
Standing alert	1.25 ± 0.5	3.0 ± 0.87	1.9	n.s
Self-grooming	0.8 ± 0.2	1.25 ± 0.4	0.7	n.s

Results based on the same individuals in a paired t-test (P= two tailed), n=9.

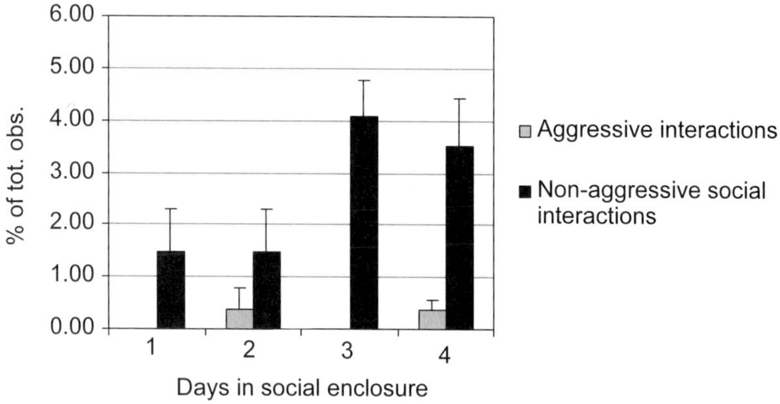

Figure 1. Social behaviours (mean ± SE %) in the joint enclosure over time.

Chasing between horses was observed mainly at the start of the social period and each chasing-episode never lasted more than 4 minutes at a time. The horses ran in relation to such chasing-episodes and the amount of running decreased during the period in social enclosure (Day 1: 3.3 ± 0.9% versus Day 4: 0.4 ± 0.2%, $P<0.01$).

Discussion

In the social enclosure the horses were more active, spending more time eating grass and less time standing passively than in the individual paddocks. This can both be due to the large enclosure size or that the horses were kept in a social group. Surprisingly, it was not the social behaviours that made up the difference in activity, but rather the fact that they increased the time spent eating grass to nearly 75%. There was, however, not much more grass to eat in the social enclosure compared to in the individual paddocks, so this finding can probably be explained by social facilitation of foraging behaviour (Sweeting *et al.*, 1985; Rifa, 1990).

Our hypothesis that the joint enclosure would entail more running and playing was not confirmed, even though the total space was much larger in this enclosure compared to the individual paddock. For Przewalski horses in social groups Hogan *et al.* (1988) and Boyd and Bandi (2002) found that the activity increased with increasing enclosure size. Correspondingly, increased space allowance for pigs resulted in higher activity (e.g. Weng *et al.*, 1998), and calves showed more play behaviour when increasing the space allowance (Jensen and Kyhn, 2000). However, space allowance is probably not the only factor that affects the horse's motivation for running and playing. For horses that are familiar with a specific paddock environment and the fact that paddocks usually are quite barren with few stimuli, the motivation for running and playing is likely to be low. Possible explanations for the low frequency of running and playing in our experiment may be the high age of the horses and that these horses had been at this farm for several years and hence were very familiar with both the paddocks and the daily activities. As commented by Fraser (1992), play behaviour mostly occur in young horses.

It is interesting to notice that the frequency of aggressive behaviours was very low. Our observation method obviously underestimated short events like aggression. Still, our results correspond well to other studies that have found few injuries and aggressive episodes in social groups of both domestic and feral horses (Christensen *et al.*, 2002 a; Grogan and McDonnell, 2005; Lehmann *et al.*, 2006). Hence, the use of larger enclosures with groups of horses may be a good alternative to individual paddocks. Generally, group housing of horses involve both lower building costs and larger total space for movement and activity (Ventorp and Michanek, 1997).

In conclusion, based on the very few incidents of aggressive interactions observed, economic benefits and the possibility for amicable social behaviours like allogrooming, the use of larger enclosures with groups of horses may be a good alternative to individual paddocks.

References

Bachmann, I. and M. Stauffacher, 2002. Housing and exploitation of horses in Switzerland: A representative analysis of the status quo. Schweiz. Arch. Tierheilkunde. 144(7): 331-347.

Boyd, L.E., 1991. The behaviour of Przewalski`s horses and its importance to their management. Appl. Anim. Behav. Sci. 29(1-4): 301-318.

Boyd, L.E., D.A. Carbonaro and K.A. Houpt, 1988. The 24-hour budget of Przewalski horses. Appl. Anim. Behav. Sci. 21(1-2): 5-17.

Boyd, L. and N. Bandi, 2002, Reintroduction of takhi, Equuas ferus przewalskii, to Hustai National Park, Mongolia: time budget and synchrony of activity pre- and post-release. Appl. Anim. Behav. Sci. 78(2-4): 87-102.

Christensen, J.W., T. Zharkikh, J. Ladewig and N. Yasinetskaya, 2002a. Social Behaviour in stallion groups (*Equus przewalskii* and *Equus caballus*) kept under natural and domestic conditions. Appl. Anim. Behav. Sci. 76: 11-20.

Christensen, J.W., J. Ladewig, E. Søndergaard and J. Malmkvist, 2002b. Effects of individual versus group stabling on behaviour in domestic stallions. Appl. Anim. Behav. Sci. 75: 233-248.

Fraser, A.F., 1992. The behaviour of the horse. 1st. ed. Wallingford, UK: CAB International, Camebridge, USA, 288 pp.

Grogan E.H. and S.M. McDonnell, 2005, Injuries and blemishes in a semi-feral herd of ponies, J. Equine Vet. Sci. 25(1): 26-30.

Hogan, E.S., K.A. Houpt and K. Sweeny, 1988. The effect of enclosure size on social interactions and daily activity patterns of the captive Asiatic wild horse (*Equus przewalskii*). Appl. Anim. Behav. Sci. 21(1-2): 147-168.

Hurnik, J.F, A.B. Webster and P.B. Siegel, 1995. Dictionary of Farm Animal Behavior, second edition, Iowa State University Press, 210 pp.

Jensen, M.B. and R. Kyhn, 2000. Play behaviour in group-housed dairy calves, the effect of space allowance. Appl. Anim. Behav. Sci. 67(1-2): 35-46.

King, S.R.B. and J. Gurnell, 2005. Habitat use and spatial dynamics of takhi introduced to Hustai National Park, Mongolia. Biol. Conservat. 124: 277-290.

King, S.R.B., 2002. Home range and habitat use of free-ranging Przewalski horses at Hustai National Park, Mongolia. Appl. Anim. Behav. Sci. 78: 103-113.

Klimov, V.V., 1988. Spatial-ethological organization of the herd of Przewalski horses (*Equus przewalskii*) in Askania-Nova. Appl. Anim. Behav. Sci. 21: 99-115.

Lehmann, K., E. Kallweit and F. Ellendorff, 2006. Social hierarchy in exercised and untrained group-housed horses--A brief report. Appl. Anim. Behav. Sci. 96(3-4): 343-347.

Miller, R., 1983. Seasonal movements and home ranges of feral horse bands in Wyoming`s Red desert. J. Range Manag. 36(2): 199-201.

Rifa, H., 1990. Social facilitation in the horse (Equus caballus). Appl. Anim. Behav. Sci. 25(1): 167-176.

Statens Jordbruksverk, 2004. Ny hästräkning genomførd. Pressemelding, Sverige, Available online:http://www.sjv.se/startsida/pressekontakten/pressemeddelanden/pressemeddelanden/5.12022b71008e0e5daa800025877.html [04/18, 2005].

Strand, S.C., S. Tiefenbacher, M. Haskell, T. Hosmer, S.M. McDonnell, and D.A. Freeman, 2002. Behaviour and physiologic responses of mares to short-term isolation. Appl. Anim. Behav. Sci. 78(2-4): 145-157.

Sweeting, M.P., C.E. Houpt and K.A. Houpt, 1985. Social Facilitation of feeding and time budgets in stabled ponies. J. Anim. Sci. 60(2): 369-374.

Søndergaard, E., E. Clausen, J.W. Christensen and H. Schougaard, 2002. DJF rapport nr 39. Oppstalding og hold af heste. Danske anbefalinger. Danmarks Jordbrugs Forskning, Tjele, Denmark, 105 pp.

Ventorp, M. And P. Michanek, 1997. Att bygga Häststall: en idèhandbok. The Agricultural University of Sweeden, Lund, Sweeden. 432 pp.

Weng, R.C., S.A. Edwards and P.R. English, 1998. Behaviour, social interactions and lesion scores of group-housed sows in relation to floor space allowance. Appl. Anim. Behav. Sci. 59(4): 307-316.

Part D

Equine behaviour welfare

The management of horse behaviour in captivity and its welfare implications

D.S. Mills
Animal Behaviour, Cognition & Welfare Group, University of Lincoln, Dept of Biological Sciences, Riseholme Park, Lincoln LN2 2LG, United Kingdom

Abstract

Recently there has been a growth in the scientific study of the welfare of horses, with behaviour commonly being used as a tool, to assess the well-being of captive equids and the effect of management interventions. An essential prerequisite to this is an appreciation of the welfare consequences of different forms of behaviour and the methods used for their manipulation. In some instances the behaviour of a horse may not relate in any significant way to a horse's well-being, whilst in other circumstances the behaviour itself might cause harm, or it might directly reflect a poor welfare state, finally behaviour may be an attempt to reduce suffering. When any of these responses are problematic to the owners or managers, how they are managed also has obvious welfare implications. Not only is the relationship between the behaviour (before and after intervention) and welfare of importance, but also the impact on welfare of the type of strategy chosen. Broadly speaking management might aim to prevent a response, address the causal factors underlying the behaviour (either through the removal of causal stimuli or by altering the perception of these), redirect the response towards a more acceptable substrate or seek to encourage other behaviours in place of the problematic behaviour. It is possible to make some generalisations about the welfare implications of each of these strategies, which should be considered when attempting to change the behaviour of a horse in captivity, in addition to the specific risks of a potential method.

Keywords: behaviour, horse, management, suffering, welfare

Introduction

There are many reasons to be interested in horse welfare. From an ethical perspective it might be argued that we have a responsibility to minimise the suffering we cause to other animals with whom we interact either directly or indirectly. From a practical point of view, animals with good welfare might be expected to perform their work more efficiently. From an academic perspective the assessment of animal welfare is also a challenging intellectual task. However, it is not the aim of this paper to discuss why we should measure horse welfare or the ethics of what we do to horses, rather it is the aim of this paper to review how the behaviour of horses in captivity relates to its welfare and how our management of horses can affect this. One of the problems with writing about animal welfare is deciding exactly what is meant by the term 'welfare'. There is no consensus definition of welfare even amongst welfare scientists. For some, welfare equates with the feelings of an individual at a given time. For others this might be referred to as its well-being, with welfare considering not just the animal's current feelings but also the threats to its well-being. Thus it might be argued that horses whose work puts them at risk of particular injury, such as limb injury in race horses, have worse welfare than those whose work does not pose such risks. This approach can be useful when considering welfare within the context of populations, but it is important not to forget the individual. Even if the risk were 1 in 10,000 for a catastrophic injury, the consequences for that one individual that is injured are disastrous. Some also consider the effort required by an animal to maintain its state of well-being. Thus Broom (1986) defines welfare as the state of an animal with regards to its attempts to cope with its environment. This would imply that a horse who is kept in an optimal ambient temperature may be thought to have better welfare than one who is having to devote additional resources to the maintenance of such a temperature. It is therefore important to clarify what an individual means by the term 'welfare' especially when one animal's welfare is said to be better than anothers. In this paper a definition based on the feelings of the animal will be adopted.

Whilst it is reasonable to assume that a horse is adapted to its natural environment, the occurrence of normal behaviour patterns does not necessarily imply good welfare. Some behaviours are clearly associated with aversion, such as flight from a predator and so their occurrence is undesirable. In other instances the significance of the behaviour depends on the context. For example, horses may move into water away from dry land in order to avoid the effects of blood-sucking insects which can cause anaemia. Whilst this is obviously beneficial to the horse, this behaviour is not without its costs. They may reduce grazing or reduce other important behaviours as a trade off for escaping the insects. So, whilst standing idle or engaging in social exchanges with others does not appear to be a behaviour of concern, in this case the horses would undoubtedly be better off doing something else if the insects were absent.

Behaviour is a form of adaptation to the environment and so is meaningless without a context. Two horses in different environments might be expected to show different behaviours as well as different patterns of behaviour as part of their healthy adaptation to the different environments. So there is no norm against which the behaviour of animals can be meaningfully assessed. Instead, individuals may be tempted to make arbitrary comparisons when it suits their case, for example the amount of time a horse should spend grazing or alert in a day to be normal or psychological healthy. There is no logic in the assumption that quantitative or qualitative differences in behaviour *necessarily* imply a difference in welfare. Instead it is important to appreciate the function and regulation of each behaviour in the context in which it occurs. Understanding the natural behaviour of horses is important to the scientific study of equine welfare as it is through this we can appreciate the functional significance of a given behaviour and the mechanisms which may underlie the adaptability of the horse and when these may be strained in an unnatural setting, for example concentrate feeding or the use of raised hay nets.

Horses have evolved a range of adaptations to help them to cope with the challenges they encounter in their natural environment. Some of these processes are useful in helping the horse to cope with challenges unique to the domestic environment, but in other cases the horse may not be able to adapt adequately, in which case there must be concern for the animal's welfare.

Riding horses are usually housed in conditions which differ from those in which their ancestral precursors evolved. Horses naturally live in small stable social groups (bands), with several bands in a given area forming a larger structured social unit often referred to as a herd (Miller and Denniston, 1979). Unlike many other domesticated mammal species, the male stays with females throughout the year in the wild state (Clutton-Brock, 1989). Within a band there are also strong bonds between specific individuals which may last a lifetime (Waring, 2003). When given access to pasture, horses spend about two thirds of the day grazing. In this time it has been calculated that horses take around 30,000 bites of grass (Mayes and Duncan, 1986) and chew around 57,600 times (Cuddeford, 1999). During grazing, horses normally feed at a given area for a few bites before walking to a new area. It has been estimated (Houpt, personal communication) that during this process alone, a horse will take approximately 10,000 steps a day. However, in the captive domestic situation prevalent in many European countries, most horses are kept individually housed in boxes around 3.5m by 3.5m with restricted social contact, limited turnout in changeable groups and a controlled feeding regime, based on rationed forage supplemented by high energy concentrate meals (Mills and Clarke, 2002). Stallions are usually kept isolated from mares and breeding may only be allowed under close supervision with the minimum of contact necessary. When there is such a mismatch between an animal's evolutionary and captive environments and the adaptations needed to survive in each, there are solid grounds for concern over the challenges faced by the occupants and their consequent welfare (Broom and Johnson, 1993; Fraser *et al.*, 1997). It is not surprising that the behaviour of horses in such domestic circumstances differs considerably from that observed in their wild or feral conspecifics (Kiley Worthington, 1987; Mills and Clarke, 2002). These differences may be quantitative (i.e. relate to the amount of time spent in each behaviour) and/or qualitatitive (i.e. relate to behaviours which are not seen in the wild/ feral state). Some of the behaviours observed in horses in this sort of housing are considered undesirable, either because they reduce the animal's quality of life, or because they appear unsightly, offensive or incomprehensible. These represent different types of problem and elimination of the behaviour does not necessarily imply resolution of any related welfare problem.

Behavioural modification is commonly undertaken either to improve welfare, or for our convenience, or to allow more aesthetically pleasing natural behaviour. A number of approaches can be used including: prevention of the behaviour itself; removal of its causal factors; redirection of the activity to more desirable substrates; and encouraging other behaviours at this time (increasing behavioural competition). This paper considers the implications of an animal's behaviour for its welfare and the theoretical framework for assessing the effect of modification on both reducing undesirable behaviour and on quality of life. An understanding of these principles is vital when considering humane treatment options for problem behaviour in horses.

Behaviour and suffering in horses

Inherent within the theory of evolution is the concept of optimality. A well adapted animal performs more efficiently in its environment than one that is less well adapted to that environment and so leaves more offspring similarly adapted. Perhaps because of the special role of the veterinary profession in protecting animal welfare, measures of health, disease and physiological function have been the primary measures of welfare used. However, it must be remembered that whilst good welfare requires good health, good health does not always mean good welfare as other problems may exist. Physical health measures should not be the only indices of welfare although they can be useful in flagging up areas of major concern, e.g. the prevalence of chronic obstructive pulmonary disease in housed horses. However, it is also possible for medicine to mask welfare problems. For example, potentially lame animals can be given analgesics to allow them to train and compete in the short-term at potentially enormous cost in the longer-term. Thus it is important to remember the scientific axiom: absence of evidence does not equate with evidence of absence. Suffering can be thought of as a range of aversive states, such as fear, pain and frustration that are subjectively experienced (Dawkins, 1990). These evolved as adaptive responses to environmental challenges to maximise reproductive fitness, to avoid injury and to minimise the risk of death. Whilst suffering is a concept that cannot yet be directly assessed or precisely measured in animals, it can be inferred from their physiological and behavioural responses (Dawkins, 1980). An animal's behaviour can relate to welfare in four ways; firstly it can have no implications for welfare; secondly it can cause harm and suffering in the performer or in others; thirdly it can be an indicator of poor welfare; and fourthly it can be a means of reducing suffering.

Behaviour which is of no welfare significance

When an animal behaves in a given way, it does so at the expense of other behaviours which could be expressed at that time. The brain constantly balances the motivational inputs from both external and internal stimuli, in order to select the most appropriate action at any given time (Norman and Shallice, 1986). Thus behaviours compete for expression, through self-reinforcement and reciprocal inhibition processes. The decision as to which behaviour to perform at any given time is based on short-term or proximate rules of thumb (McFarland, 1989) that over evolutionary time scales have been selected because they approximate to optimal behaviour (Krebs and Davies, 1987). However, during the decision making process, the animal can not know the ultimate consequences of every action and in captivity, animals will adhere to these rules of thumb, even though the long term evolutionary consequences no longer hold. As a result, they may behave in a manner that appears to be dysfunctional, incomprehensible, unnatural, unattractive or even 'stupid'. This can occur because motivationally significant cues are encountered out of context. Robins, for example, perform aggressive territorial behaviour which can be initiated simply by presenting the robin with a tuft of red feathers; sticklebacks respond aggressively to models with red markings that bear a crude resemblance to rival male sticklebacks (Manning and Dawkins, 1992); and well fed, indoor cats may pounce on moving shiny sweet wrappers. In these situations the animal is responding to key features of its environment, which in the wild would be biologically meaningful, but can be manipulated in captivity to elicit apparently pointless responses. The extent to which this occurs in captive horses is unknown but deserves investigation.

Alternatively unusual patterns of behaviour can develop because the reinforcing properties of the captive environment differ from those in the wild. Skinner (1938) trained rats to press levers to obtain food within an artificial environment, where provision of environmental reinforcement was instrumental on the performance of novel or arbitrary responses. In this way, bizarre behaviours such as unusual ingestive postures by horses and possibly some pre-feeding rituals can be learnt through trial and error and exhibited without illustrating anything about the animal's well-being. So behaviours that are unique to captivity are not necessarily indicative of suffering or a problem in the environment, but may simply be an alternative means of expression of the animal's behavioural repertoire. In these cases, any intervention should be justified on some other ethically sound basis.

Behaviour which causes harm and suffering

Many activities found in captive animals are harmful either to the performer or to other animals. Self-injury, is commonly recognised as a stallion behaviour problem (Dodman *et al.*, 1988; Houpt, 1983) but may occur in any sex and in immature individuals (Dodman *et al.*, 1994). Aggressiveness is also one of the most common problems in the horse which is referred for treatment (Houpt, 1981).

When a pattern of behaviour is overtly harmful then there is a strong case for intervention for the benefit of those who may be injured. However, it is wrong to assume that intervention is imperative in these cases. The potential benefit to be gained versus the potential cost of non-intervention must be considered. The potential benefit depends not only on the prognosis for success, but also the welfare cost of the method used to achieve this (see below).

Behaviour which is indicative of poor welfare

It might be thought that all apparently functionless behaviours in restrictive conditions are context specific equivalents of natural behaviours in extensive systems and consequently unrelated to welfare. However, this is unsound; firstly, because both behaviours may be a response to a common noxious stimulus and secondly because their performances may differ in how effectively they address the underlying motivation. Activities such as limping on a painful leg (Adams, 1974) or avoiding aversive situations are overtly functional responses which are clearly related to a noxious stimulus. Pain, however, is not a simple concept. Pain is defined as: 'An unpleasant sensory and emotional experience associated with actual or potential tissue damage or described in terms of such damage' (IASP, 1979). The recognition of pain has received most scientific study, perhaps because of the historical importance of the veterinary profession in the study of animal welfare and its obvious association with physical lesions. A variety of techniques have been proposed in the scientific literature for the assessment of pain in the horse based on behavioural and physiological measures e.g. the assessment of activity budgets pre and post surgery (Price *et al.*, 2003), response to analgesia (Dyson 2002), and median frequency within electro-encephalogram recordings (Murrell *et al.*, 2003), but these are generally restricted to use within a specific context. More general measures, such as indicators of sympathetic nervous system activity tend to be non-specific to suffering, unvalidated and / or contradictory. Nonetheless, the ethology of pain is probably worthy of further attention.

But suffering is not just about pain, many apparently functionless responses observed in captivity may also relate to noxious challenges. Dawkins (1990) argues that suffering consists of a 'a wide range of emotional states that occur when an animal is blocked from carrying out actions that are biologically mandated, normally reduce harm or risk to life or concern reproduction.' This stance has spawned a wealth of work designed to assess the needs of animals in order to determine which are biologically mandated. Simple preference tests may tell us what an animal prefers but they do not tell us if an animal is suffering if it is deprived of the preferred choice. I may prefer Bordeaux wines to Champagnes but I can be happy with either! So scientists have developed techniques where they have started to look at the price an animal is willing to pay for a given commodity. In this way we can have a clearer idea of what is really important to an animal. However, whilst there is nothing to theoretically stop such work being done in the horse, the cost of building the experimental apparatus and housing the number of animals necessary for the time required to obtain sound data has to date largely prevented such work in the horse; although Houpt's group at Cornell (Lee *et al.*, 2001) has conducted preliminary studies to assess the strength of a horse's motivation for exercise and companionship when confined for 23 hours in the day. A higher price was paid for companionship over exercise and this reinforces much work by our own group which suggests that social isolation is one of the primary problems with many modern housing systems (see Mills and Clarke, 2002 for a review). There is undoubtedly a need for more work in this field and we can only hope that those who have the potential to fund it recognise its importance so we can objectively assess the welfare of the horse in a variety of contexts. Simple observations of behaviour may provide a useful background and theoretical basis for such work. Thus a higher incidence of woodchewing is reported in horses at times when it might be expected that the horse's stomach is less full than would occur in the natural state (Krzak *et al.*, 1991), thus this may be an expression of high feeding motivation, and as such an indicator of hunger.

The responses used by animals in frustrating circumstances may differ in how well they satisfy their motivational origin. Little research has been carried out to establish this in the horse and it would be useful to investigate the degree to which unnatural and natural substrates are substitutes for one another. However, Mills and Davenport (2002) have shown that when a mirror is introduced into the stable of a horse it appears to have the same effect as allowing social contact through a small window between stables. Further research is also required into the potential psychological heterogeneity which may underlie physically similar behaviours (Mason, 1991). For example, repetitive behaviour problems may reflect an underlying state of anxiety (Rapaport, 1988) or psychotic disturbance (Astrachan *et al.*, 1972), alternatively there may be no awareness of the behaviour or its impact on those around the performer and so of no relevance to the well-being of the performer (American Psychiatric Association, 1994).

Behaviour which helps to reduce suffering

There has been much work on whether an animal's behaviour in captivity can include elements which function as a means of coping with an adverse environment. It has been suggested that crib-biting in horses, may have a role in stress modulation since the performance of the behaviour is associated with lowered heart-rates and thermal threshold for limb withdrawal (Lebelt *et al.*, 1998) and reduced blood corticosteroid levels (McGreevy and Nicol, 1998b). The development and initiation of these behaviours are often associated with arousing situations, but it is not clear as to whether their performance directly affects physiological measures of arousal i.e. it is a genuine coping strategy, or if the effect is co-incidental (Dantzer, 1991). More recently, Nicol (1999) has suggested that cribbing may help to reduce suffering, by producing alkaline saliva to buffer against the excessive gastric acidity caused by feeding horses a small number of concentrated meals. However, this hypothesis requires careful investigation since recent work by McCall (Houpt personal communication) suggests that during cribbing acidity may increase in the stomach. Nonetheless Mills and Macleod (2002) found that an antacid formulation helped to reduce cribbing in horses around the time of feeding.

Behaviour and suffering: summary

An animal's behaviour is the product of many decision making processes and a particular response can result from a number of mechanisms that do not necessarily involve either conscious or emotional processes. This lack of precise measures for suffering does not, however, imply that animals are incapable of suffering. Animals have evolved similar mechanisms to those of humans to control their behaviour and this probably extends to the emotional responses (Dawkins, 1990). Whilst there may be difficulties in the precise determination of the cognitive processes involved in any given situation, there is nevertheless strong evidence for relationships between behaviour and well-being in captivity from both physiological and motivational studies. It is vital to understand the causes and effects of specific activities before attempting modification so that undesirable behaviour can be treated without harming the animal's quality of life. The problem behaviour may be an adaptive response to captivity which reduces or prevents suffering, so preventing the behaviour will only remove the behavioural signs of a poor environment, but not cure the underlying welfare problem. Alternatively intervention may reduce the animal's ability to cope with its environment, so treatment by preventative means can carry a considerable risk of reducing welfare by disabling the animal's adaptive responses to captivity.

The modification of behaviour in horses

Behaviour modification can adopt a number of approaches. It can aim to:
1. prevent the response;
2. to address the causal factors underlying the behaviour;
3. redirect the response towards a more acceptable substrate;
4. encourage other behaviours (increase behavioural competition) with a view to altering the tendency for the behaviour.

These may be achieved by environmental manipulation (e.g. enrichment), chemical (e.g. dietary or pharmacological), physical (e.g. surgical) intervention or behaviour conditioning (e.g. training). These do not represent the only means for resolving behaviour problems since problems may also be eliminated through counselling designed to bring about changes in the owner's perception of the behaviour. In these latter circumstances the welfare implications of treatment relate to the welfare of the ongoing behaviour and its likely future development, since treatment brings about no change in the behaviour.

Prevention of the behaviour

Prevention can be achieved in a number of ways including the use of: physical barriers such as weaving bars above stable doors (McBride and Cuddeford, 2001), and collars that restrict crib-biting or wind-sucking (McGreevy and Nicol, 1998a), psychological barriers such as punishment using electric shock have also been advocated for the control of crib-biting in horses (Baker and Kear-Colwell, 1974); pharmacological intervention such as arousal reduction with sedatives like acepromazine (Nurnberg *et al.*, 1997); environmental manipulation aimed at eliminating consummatory substrates, such as the horizontal surfaces used by cribbing horses (McGreevy and Nicol, 1998c); and surgical intervention, such as myectomy (Forsell, 1926) or combined neurectomy and myectomy (Greet, 1982) for cribbing and windsucking.

The prevention of behaviour alone may cause distress to the animal. Even apparently innocuous preventative measures such as the removal of cribbing surfaces raises blood corticosteroid concentration, unless access is given to hay (McGreevy and Nicol, 1998c). It is, therefore, important to have good cause to employ these measures, and, more often than not, preventative measures are employed for cultural or aesthetic reasons rather than the benefit of the animal. Even if there is good reason to modify the behaviour, then prevention may not be the best approach, because the behaviour may be symptomatic of a poor environment or may have adaptive value in artificial or unusual conditions. Preventative measures do not address the underlying causes of the behaviour, so the animal may still be highly motivated to persevere with the response in a reduced, alternative or redirected form. Weaving bars, for example, reduce head-out-of-the-door weaving, but horses can continue to weave within the stable (McBride, 1996). Surgical interference to prevent repetitive behaviour problems, is commonly recommended within the veterinary literature, but is difficult to justify on welfare grounds except where it is reducing suffering, such as the pain that might exist in certain forms of headshaking (Cook, 1979a; b; 1980a; b;1992; Mills *et al.*, 2002).

Intervention on underlying causes

Behaviour modification can be achieved by addressing the causal factors, either through the removal of causal stimuli or by manipulation of endogenous factors that link stimulus and response, so the animal is no longer motivated to behave inappropriately. The causal mechanisms that link stimuli in the external and internal environments with the animal's response can be manipulated in a number of ways such as through: psychological intervention e.g. by using controlled incremental exposure and continuous exposure techniques such as systematic desensitisation (Wolpe, 1958) or behavioural flooding (Baum, 1988); or surgical intervention such as orchidectomy for stallion-like behaviour in a rig (supposedly castrated horse with one or two retained testicles). Neurectomy of the infra-orbital nerve is proposed in some cases to reduce the transmission of noxious sensations from the muzzle which trigger headshaking (Mair, 1999). The potential benefit of such a procedure should be considered carefully against the concerns it raises for the welfare of the animal involved. Surgery through the pain it inevitably causes involves poor welfare before the goal is achieved and is associated with a range of potential complications (Mair and Lane, 1990; Mair, 1999). The long term effects of sensory deprivation may be distressing and reduce the horse's general ability to respond to its surroundings appropriately, resulting in serious self-trauma. The relationship between pharmacological interventions and causal mechanisms is unclear and a topic of debate within the veterinary profession. The factors involved are complex and should not be trivialised through simple generalisation; the reader is referred to Mills (2003) for a discussion of these matters.

The removal of what appear to be the environmental causes of the behaviour can fail to treat an activity. Horses who wind-suck or crib-bite in the stable may persist with the behaviour at pasture (Kiley-Worthington, 1977). This could be because the environmental change does not actually solve the motivational problem, alternatively, the behaviour may have become independent of its original causes or external cues in general. In voles, providing natural cover is less effective at reducing stereotypy as they age (Cooper *et al.*, 1996). This can occur either because the behaviour is self reinforcing or the animal has become generally less responsive to external cues (Fentress, 1976; Cooper and Nicol, 1991). However, it is unclear whether such a mechanistic change occurs in equine repetitive behaviour problems (McAfee *et al.*, 2002).

Where removal of the putative environmental causes does not reduce an undesirable behaviour, where the causal factors are unknown or where removal of causal factors is inconvenient, then diversion of the response on to a more desirable substrate; or increasing behavioural competition can treat undesirable behaviour without the welfare risks of attempts at simple prevention of the behaviour.

Redirection of the response

Redirection of undesirable behaviour to less harmful, more convenient or more aesthetically pleasing substrates can be the simplest form of behavioural modification. Should crib-biting cause excessive tooth wear, covering the preferred cribbing surface with a less abrasive coating or using a tough rubber cribbing board, may be preferable to removing all surfaces (McGreevy and Nicol, 1998a). In this case, the welfare implications of the ongoing behaviour remain unchanged, but if the redirected activity causes less harm to either the performer or those around it, then it may improve the welfare of the animal's concerned.

Increasing behavioural competition

Provision of alternative resources can reduce problem behaviour in a number of other ways. For example, straw may be associated with a lower prevalence of certain behaviour problems in horses (McGreevy *et al.*, 1995) by providing behavioural or physical feedback in the form of chewing or gut-fill that reduces hunger, by allowing the redirection of behaviours which may not otherwise be expressed, or by increasing behavioural competition with straw manipulation possibly pushing problem behaviours out of the horse's time budget (Mills *et al.*, 2000). Wild animals have a limited time budget and selection will favour those who on average distribute it in an optimal way. Captive animals may have more 'spare' time (Dawkins, 1990; Mason *et al.*, 1997) as many goal-directed behaviours are satisfied by the provision of a physically optimal environment. There is a limited behavioural repertoire which may expand to fill the available time and some problem behaviours in the horse may simply represent time-filling activities (Mills and Nankervis, 1999). In these situations, it would be expected that the behaviours would respond well to techniques designed to increase behavioural competition. General changes to the animal's environment such as increased space, housing outdoors, or increased environmental complexity can reduce the incidence of undesirable behaviour (McBride and Long, 2001), not only by removing eliciting factors, or providing more appropriate means of expression, but by allowing the expression of other previously denied activities. These can then reduce the incidence of the undesirable behaviour by competing for time, attention or energy. Specific environmental enrichments for horses include the use of foraging toys like the 'Equiball' (Henderson and Waran, 2001), which may act through this mechanism. However, it is important that these devices do not cause further frustration and associated problems. Responses which compete for expression with a problem behaviour in response to the causal stimuli may be deliberately trained as occurs with counter-conditioning (Mills and Nankervis, 1999).

Conclusion

Understanding why a specific activity is performed and the consequences of treatment depends on identifying: the causal factors in the environment; costs and benefits of the activity and its treatment; whether or not the activity is truly undesirable and to whom. Only then can the condition be effectively treated confidently with a view to optimising the animal's welfare. Many responses to captivity that undergo treatment cause no direct harm to the animal and are frequently perceived as undesirable on aesthetic, convenience and economic grounds. Before treatment is proposed the therapist must evaluate the risks of frustrating an established behaviour, taking away the animal's means of adapting to an unfavourable environment, or even disabling its stress modulatory mechanisms altogether. Thus surgical prevention of cribbing, as has been frequently advocated by the veterinary profession (Dalacalle *et al.*, 2002; Firth 1980; Forssel, 1926; Fraufelder, 1981; Hamm, 1977; Hakansson *et al.*, 1992; Karlander *et al.*, 1965; Smith, 1924) may not just be unjustifiable but actually contra-indicated on welfare grounds. It is often preferable to consider treatments which remove the underlying causes in the environment or deliberately redirect the behaviour to a more desirable form. These carry lower risks of adversely affecting welfare, and reduce the tendency to persevere with the response or express it in a redirected, but still undesirable form.

In situations where the behaviour appears to be independent of external causal factors, where the causal factors are unknown or where their removal is impractical, then there may be some justification in directly curbing the behaviour itself. This is most likely when the behaviour causes harm. In these circumstances the benefits to the animals should outweigh the costs of the treatment. Nevertheless, attempts to reduce the undesirability of the animal's behaviour should primarily focus on positive welfare techniques such as increasing behavioural competition by allowing alternative activities, or by redirecting the undesirable response to more appropriate substrates. However, there is a need for further research which evaluates the effect of enabling horses to express a greater variety of behaviours within the constraints of the stable environment, in order to be confident about the actual welfare outcome of these techniques. Where the behaviour causes no harm to the animal and there is a risk of reducing its welfare using preventative measures, it is preferable to educate the owner as to the costs and benefits of intervention. This is, in itself, a valuable treatment of 'the problem'.

A common criticism of those working in the field of animal welfare is that we cannot know the mind of another. But this criticism also reflects a failure to understand the philosophy of science. According to the scientific method, a theory or proposition is strengthened by the elimination of competing hypotheses, but can never be validated. All scientific evidence carries with it a degree of uncertainty and we decide what level we are prepared to accept (often a 5% chance level). However these principles are often forgotten by sceptics who, for whatever reason, appear not to want to contemplate what might be happening in the minds of other animals. There is a need for those interested in animal welfare to understand the philosophy of science in order to appreciate what is possible, realistic and reasonably justifiable. Most progress will be made with an interdisciplinary approach which seeks to identify where problems may arise and how they may be managed most humanely. In this way the well-meaning intentions of owners will translate into the well-being of their charges.

Acknowledgements

I am most grateful to Jonathan Cooper for his input and discussion on an earlier draft of this work.

References

Adams, R.O., 1974. Lameness in horses (3rd edn). Lea and Febiger, Philadelphia, USA.

American Psychiatric Association, 1994. Diagnostic and Statistical Manual of Mental Disorders. (4th edn). American Psychiatric Association, Washington, USA.

Astrachen, B.M., M. Harrow, D. Adler, L. Brauer, C. Schwartz and G. Tucker, 1972. A checklist for the diagnosis of schizophrenia. British J. Psychiatry 121: 529-539.

Baker, G.J. and J. Kear-Colwell, 1974. Aerophagia (wind-sucking) and aversion therapy in the horse. Proceedings of the American Association of Equine Practitioners 20: 127-130.

Baum, M., 1988. Contributions of animal studies of response prevention (flooding) to human exposure therapy. Psychol. Reports 63: 421-442.

Broom, D.M., 1986. Indicators of poor welfare. British Vet. J. 142: 524-526.

Broom, D.M. and K.G. Johnson, 1993. Stress and Animal Welfare, Chapman and Hall, London, U.K.

Clutton-Brock, T.H., 1989. Mammalian mating systems. Proc. Royal Soc. B, 236: 339-372.

Cook, W.R., 1979a. Headshaking in horses: part 1. Equine Pract. 1: 9-17.

Cook, W.R., 1979b. Headshaking in horses: part 2. History and management. Equine Pract. 1: 36-39.

Cook, W.R., 1980a. Headshaking in horses: part 3. Diagnostic tests. Equine Pract. 2: 31-40.

Cook, W.R., 1980b. Headshaking in horses: part 4. Special diagnostic procedures. Equine Pract. 2: 7-15.

Cook, W.R., 1992. Headshaking in horses: an afterword. Compendium Cont. Ed. Pract. Vet. Surgeon 14: 1369-1372.

Cooper, J.J. and C.J. Nicol, 1991. Stereotypic behaviour affects environmental preference in bank voles. Anim. Behav. 41: 971-977.

Cooper, J.J., F.O. Odberg and C.J. Nicol, 1996. Limitations on the effect of environmental improvement in reducing stereotypic behaviour in bank voles (*Clethrionomys glareolus*). Appl. Anim. Behav. Sci. 48: 237-248.

Cuddeford, D., 1999 Why feed fibre to the performance horse today? Proceedings of the British Equine Veterinary Association Specialist Days on Behaviour and Nutrition. P.A. Harris, G.M. Gomarsall, H.P.B. Davidson and R.E. Green (Eds.) Equine Veterinary Journal Ltd, Newmarket, U.K. 50-54.

Dantzer, R., 1991. Behavioural, physiological and functional aspects of stereotypic behaviour: a review and reinterpretation. J. Anim. Sci. 62: 1776-1786.

Dawkins, M.S., 1980. Animal Suffering; The Science of Animal Welfare. Chapman and Hall, London, U.K.

Dawkins, M.S., 1990. From an animal's point of view; Motivation, fitness and animal welfare. Behav. Brain Sci. 13: 1-61.

Delacalle, J., D.J. Burba, J. Tetens and R.M. Moore, 2002. Nd: YAG laser-assisted modified Forsell's procedure for treatment of cribbing (crib-biting) in horses. Vet. Surgery 31: 111-116.

Dodman, N.H., L. Shuster, M.H. Court and J. Patel, 1988. Use of a narcotic antagonist (nalmefene) to suppress self-mutilative behavior in a stallion. J. Am. Vet. Med. Assoc. 192: 1585-1586.

Dodman, N.H., J.E. Normile, L. Shuster and W. Rand, 1994. Equine self-mutilation syndrome (57 cases). J. Am. Vet. Med. Assoc. 204: 1219-1223.

Dyson, S., 2002. Subjective and quantitative scintigraphic assessment of the equine foot and its relationship with foot pain. Equine Vet. J. 34: 164-170.

Fentress, J.C., 1976. Dynamic boundaries of patterned behaviour: interaction and self-organisation. Growing Points in Ethology. P.P.G. Bateson and R.A. Hinde (Eds.), Cambridge University Press, Cambridge, U.K. 135-169.

Firth, E.C., 1980. Bilateral ventral accessory neurectomy in windsucking horses. Vet. Rec. 106: 30-32.

Forsell, G., 1926. The new surgical treatment against crib-biting. Vet. J. 82: 538-548.

Fraser, D., D.M. Weary, E.A. Pajor and B.N. Milligan, 1997. A scientific conception of animal welfare that reflects ethical concerns. Anim. Welf. 6: 187-205.

Fraufelder, H., 1981. Treatment of crib-biting: A surgical approach in the standing horse. Equine Vet. J. 13: 62-63.

Greet, T.R.C., 1982. Windsucking treated by myectomy and neurectomy. Equine Vet. J.14: 299-301.

Hakansson, A., P. Franzen and H. Petersson, 1992. Comparison of two surgical methods for treatment of crib-biting in horses. Equine Vet. J. 24: 494-496.

Hamm, D., 1977. A new surgical procedure to control crib-biting. Proc. Am. Assoc. Equine Pract. 23: 301-302.

Henderson, J.V. and N.K. Waran, 2001. Reducing equine stereotypies using the Equiball ™. Anim. Welf. 10: 73-80.

Houpt K.A., 1981. Equine behavior problems in relation to humane management. Int. J. Study Animal Problems. 2: 329-337.

Houpt, K.A., 1983. Self-directed aggression: a stallion behavior problem. Equine Pract. 5: 6-8.

International Association for the Study of Pain. (IASP), 1979. Pain terms: a list with definitions and notes on usage. Pain 6: 249-252.

Karlander, S., J. Mansson and G. Tufvesson, 1965. Buccostomy as a method of treatment for aerophagia (wind-sucking) in the horse. Nordisk Veterinär Medicin 17: 455-458.

Kiley-Worthington, M., 1977. Behavioural Problems of Farm Animals. Oriel Press, Stocksfield, U.K.

Kiley-Worthington, M., 1987. The Behaviour of Horses in Relation to Management and Training. J.A. Allen, London, U.K.

Krebs, J. and N. Davies, 1987. Behavioural Ecology: An Evolutionary Approach. Blackwell Scientific, Oxford, U.K.

Krzak, W.E., H.W. Gonyou. and L.M. Lawrence, 1991. Wood chewing by stabled horses: diurnal pattern and effects of exercise. J. Anim. Sci. 69: 1053-1058.

Lee, J., T. Floyd and K. Houpt, 2001. Operant and two choice preference applied to equine welfare. In: Proc. 35th Int. Congress ISAE, Garner J.P., J.A. Mench, S.P. Heekin (Eds.), The Centre for Animal Welfare, Davis, USA, 110.

Lebelt, D., A.J. Zanella and J. Unshelm, 1998. Physiological correlates associated with cribbing behaviour in horses; changes in thermal threshold, heart rate, plasma B-endorphin and serotonin. Equine Vet. J. Supplement 27: 21-27.

Mair, T.S., 1999. Assessment of bilateral infra-orbital nerve blockade and bilateral infra-orbital neurectomy in the investigation and treatment of idiopathic headshaking. Equine Vet. J. 31: 262-264.

Mair, T.S. and J.G. Lane, 1990. Headshaking in horses. In Practice 9: 183-186.

Manning A. and M.S. Dawkins, 1992. An Introduction to Animal Behaviour. Cambridge University Press, Cambridge, U.K.

Mason, G.J., 1991. Stereotypies: a critical review. Anim. Behav. 41: 1015-1038.

Mason, G.J., J.J. Cooper and J. Garner, 1997. Models of motivational decision making and how they affect the experimental assessment of motivational priorities. Animal Choices. Forbes, J.M., T.L.J. Lawrence, R.G Rodway and M.A Varley (Eds.), BSAS, Penicuik, Scotland. 9-17.

Mayes, E. and P. Duncan, 1986. Temporal patterns of feeding behaviour in free-ranging horses. Behaviour 96: 105-129.

McAfee, L.M, D.S. Mills and J.J. Cooper, 2002. The use of mirrors for the control of stereotypic weaving behaviour in the stabled horse. Appl. Anim. Behav. Sci. 78: 159-173.

McBride, S.D., 1996. A comparison of physical and pharmacological treatments for stereotypic behaviour in the horse. Proc. 30th Int. Congress ISAE, Duncan, I.J.H., T.M. Widowski and D.B. Haley (Eds.), Centre for the Study of Animal Welfare, Guelph, Canada. 26.

McBride, S.D. and D. Cuddeford, 2001. The putative welfare-reducing effects of preventing equine stereotypic behaviour. Anim. Welf. 10: 173-189.

McBride, S.D and L. Long, 2001. Management of horses showing stereotypic behaviour, owner perception and the implications for welfare. Vet. Record 148: 799-802.

McFarland, D., 1989 Problems in Animal Behaviour. Longman Scientific, London, U.K.

McGreevy, P.D., P.J. Cripps, N.P. French, L.E. Green and C.J. Nicol, 1995. Management factors associated with stereotypic and redirected behaviour in the Thoroughbred horse. Equine Vet. J. 27: 86-91.

McGreevy, P.D. and C.J. Nicol, 1998a. Prevention of crib-biting: a review. Equine Vet. J. Supplement 27: 35-38.

McGreevy, P.D. and C.J. Nicol, 1998b. Physiological and behavioural consequences associated with short-term prevention of crib-biting in horses. Physiol. Behav. 65: 15-23.

McGreevy, P.D. and C.J. Nicol, 1998c. The effect of short term prevention on the subsequent rate of crib-biting in Thoroughbred horses. Equine Vet. J. Supplement 27: 30-34.

Miller, R. and R.H. Denniston, 1979. Interband dominance in feral horses. Z. tierpschol. 51: 41-47.

Mills, D.S., 2003. Medical paradigms for the study of problem behaviour: a critical review. Appl. Anim. Behav. Sci. 81: 265-277.

Mills, D.S. and K.J. Nankervis, 1999 Equine Behaviour: Principles and Practice. Blackwell Science, Oxford, U.K.

Mills, D.S., S. Eckley and J.J. Cooper, 2000. Thoroughbred bedding preferences, associated behaviour differences and their implications for equine welfare. Anim. Sci. 70: 95-106.

Mills, D.S. and A. Clarke, 2002. Housing, Management and Welfare. The Welfare of Horses. N.Waran (editor), Kluwer Academic, Dordrecht, The Netherlands. 77-97.

Mills, D.S., S. Cook, K. Taylor and B. Jones, 2002. Analysis of the variations in clinical signs shown by 254 cases of equine headshaking. Vet. Record 150: 236-240.

Mills, D.S. and K. Davenport, 2002. The effect of a neighbouring conspecific versus the use of a mirror for the control of stereotypic weaving behaviour in the stabled horse Anim. Sci. 74: 95-101.

Mills, D.S. and C.A. Macleod, 2002. The response of crib-biting and windsucking horses to treatment with an antacid mixture. Ippologia 13: 33-41.

Murrell, J.C., C.B. Johnson, K.L. White, P.M. Taylor, Z.L. Haberham and A.E. Waterman-Pearson, 2003. Changes in the EEG during castration in horses and ponies anaesthetized with halothane. Vet. Anaesthesia Analgesia 30: 138-146

Nicol, C.J., 1999. Understanding equine stereotypies. Equine Vet. J. Supplement 28: 20 -25.

Norman, D.A. and T. Shallice, 1986. Attention to action: willed and automatic control of behaviour. Consciousness and Self-regulation: Advances in Research and Theory. Davidson, R.J., G.E. Schwartz and D. Shapiro (Eds.) Plenum Press, New York, USA.1-18.

Nurnberg, H.G., S.J. Keith and D.M. Paxton, 1997. Consideration of the relevance of ethological animal models for human repetitive behavioural spectrum disorders, Biol. Psychiatry 41: 226-229.

Price, J., S. Catriona, E.M. Welsh and N.K. Waran, 2003. Preliminary evaluation of a behaviour based system for assessment of post-operative pain in horses following arthroscopic surgery. Vet. Anaesthesia Analgesia 30: 124-137

Rapoport, J.L., 1988. The biology of obsessive compulsive disorder. J. Am. Vet. Med. Assoc. 260: 2888-2890.

Skinner, B.F., 1938. The Behavior of Organisms. Appleton-Century Crofts, New York.

Smith, F., 1924. An attempt to deal with 'windsucking' by a surgical interference. Vet. J. 80: 238-240.

Waring, G., 2003. Horse Behavior. (2nd edn) Noyes, New York, USA.

Wolpe, J., 1958. Psychotherapy by Reciprocal Inhibition. Stanford University Press. Stanford, California, USA.

Impact of Nutrition and Feeding practices on equines, their behaviour and welfare

P.A. Harris
Equine Studies Group, WALTHAM Centre for Pet Nutrition, Melton Mowbray, Leics LE14 4RT, United Kingdom

Abstract

In the wild, horses spend most of the day roaming and foraging in an externally variable environment as part of a herd. As non-ruminant herbivores they are well suited to a high fibre, low starch diet. They rarely fast voluntarily for more than 2 -4 hrs at a time and would naturally forage for 16- 18hrs a day. Modern horse management often brings horses into small, enclosed, isolated environments and limits the feeding occasions. What and when they are able to eat, is now predominantly determined by ourselves and we therefore have to take responsibility for the effects that our choice of managemental practices have on their health as well as their behaviour and welfare. Many managemental practices have the potential to be less than optimal for horse welfare. In addition, many people, in developed countries, have horses in order to ride them or to watch them being ridden. Being naturally fear and flight animals, their behaviour becomes increasingly important when being ridden in environments beyond the rider's control or under competitive circumstances. Currently much of the interest around diet and nutrition therefore is focused on the areas of temperament and tractability. This paper highlights the impact that diet and the way we feed can have, both positively and negatively, on the behaviour and welfare of horses especially the young growing animal. It will concentrate on the work published by WALTHAM and its collaborators since 1995, highlighting the work on feeding practices, feeding behaviour as well as the impact of diet on behaviour.

Key words: welfare, nutrition, equine, behaviour, management

Introduction: what is the challenge?

Feeding and stable management practices vary greatly around the world and it is probable that almost any permutation of regimen could be found somewhere. Many factors influence how horses are kept, but in general, the way horses are fed and managed primarily reflects the purpose for which they are kept (e.g. native breeding pony vs., the working animal vs. racing Thoroughbred) as well as their location (third world vs. developing vs. developed: rural vs. urban); their breed/age, local or regional customs, the time of year, availability of feedstuffs etc. as well as the owner's financial situation (Harris, 1997, 2000). The potential impact of managemental practices on the welfare of the horse depends upon how the horses are being kept in the first instance. In countries where feed is limited the welfare issues tend to be different from more developed countries where problems may be more associated with an excess of feed.

Regardless, it is being increasingly apparent that many managemental practices may not be optimal for the welfare of horses given that the ideal probably includes daily access to grazing, prolonged foraging time and contact with other horses (Davidson and Harris, 2002). But, what is the ideal for the modern day horse that enables its needs to be met whilst fulfilling the requirements of its human provider? 24hr pasture turn out perhaps? However, many pastures, for climatic and geographical reasons will not provide, even if well managed, all nutritional requirements throughout the whole year especially for horses that are pregnant, lactating, growing or exercising. Overgrazing and inadequate endoparasitic control can also be detrimental to the health of horses. If left unmanaged the habit of pattern grazing can lead to the grazed areas being impoverished, bare and potentially infested with unwanted or even poisonous plants such as ragwort (*senico jacobaea*). Pastures established for other herbivores, such as cattle, may not be any better as they tend to be too 'rich' for horses. Certain animals prone to conditions such as obesity, laminitis or tying up require restricted access to pastures especially at certain times of the year. Inadequately fenced and maintained pastures may actually prove to be hazardous. Stabling, on the other hand, does provide many advantages including protection against environmental extremes and ectoparasites, provision for individual feeding or care as required, as well as reducing the risk of accidental damage by other horses. However, by confining horses to a stable we not only restrict their movement and social contact with other horses but also affect their perceived ability to remove themselves from danger (Davidson and Harris, 2002).

As the above illustrates it is not easy to provide an ideal environment for the horse. However, it is obviously important to try and therefore this review will concentrate on some of the work we have published in the areas of feeding practices, feeding behaviour and the effect of diet on behaviour.

Feeding practices

The challenge today, therefore, is not just to understand when certain practices compromise the welfare of the horses in our care but to be able to make recommendations that allow horses to remain an integral part of many people's working life or leisure time whilst minimising adverse effects on their welfare. Animal welfare has been defined ecologically as the good fit of an animal to its environment (Kronfeld *et al.*, 1998). These authors mention that this can be evaluated from the point of view of the animal and its adaptation, or alternatively from how the environment can be adapted to the animal, for example, by improved housing, training or *feeding*. Much of our work over the last few years has looked at optimal feeding practices that not only support performance but also support the physiological and psychological requirements of horses, thereby, helping to minimise any health or welfare issues. Feeding practices are important and this paper will review two key aspects - when to feed before exercise and the role of chaff addition, highlighting WALTHAM studies.

When and what to feed before exercise

There has been considerable debate across the years about when and what should be fed to horses before they are exercised and/or at a competition in order to reduce the risk of gastrointestinal disturbances and other adverse sequellae. Should they be fed or fasted and when should the hay be fed in relation to the grain and/or exercise?

Initially our studies evaluated the effect of feeding either a fat supplemented or a more traditional high carbohydrate meal and exercising 3 or 8hrs post feeding (Pagan *et al.*, 1995; Duren *et al.*, 1999). These studies showed that the most important factor influencing the glucose response during exercise was the time of feeding not the type of diet, with those horses fed the cereal based diet 3hrs before exercise demonstrating large decreases in blood glucose levels during the exercise bout. This was followed by a series of studies investigating the effect of feeding hay with or without grain (Pagan and Harris, 1999). The insulin, glucose, FFA and various haematological responses of thoroughbred horses at rest and during a simulated competition exercise test (CET: the cross-country day at a 3DE) were evaluated. Resting horses were either given hay 2hrs before, with or 4 hrs after a grain meal. Feeding hay (~ 2.3kg) before or with a grain meal (~2.3kg) significantly reduced the glycaemic response to the grain, whilst the total protein plasma concentrations and water intake both increased. It was thought that this might be as a result of an increased rate of passage of grain through the gastrointestinal tract when the hay was present. This could in turn result in reduced digestion of the grain in the small intestine especially if a large grain meal is fed, which could lead potentially to increased levels of starch reaching the hindgut causing disturbances to the microflora. Interestingly the packed cell volume decreased considerably over the first 4hrs following the grain feeding (in those fed the hay 4 hrs after the grain).

The next study evaluated the following diets in exercising horses:

1. *Fast*: overnight fast;
2. *Grain plus ad lib hay*: ad lib hay the night before and before the CET plus ~2.3kg of a grain based meal 2 hrs prior to the CET;
3. Grain *plus hay*: ~2.3kg hay 3 hrs before the CET and ~2.3kg feed 2 hrs prior to the CET;
4. *Grain*: ~2.3kg of feed 2 hr prior to the exercise with ~2.3kg hay at 10pm the night before.

Plasma volume was determined immediately before exercise using an indocyanine green clearance method. Feeding grain reduced the free fatty acid availability and, as before, increased the blood glucose disappearance during exercise. The decrease in glucose during phase A and B of the CET seen in the first study was not attenuated by feeding hay. The fasted horses had fairly stable glucose concentrations during the exercise bout but had lower post exercise glucose and insulin concentrations (Figure 1).

Feeding hay either alongside the grain or *ad libitum* resulted in reduced plasma volume and higher lactate production as well as higher heart rates during the exercise. Fasted horses had lower blood lactate after the 8-min canter compared to the others. There were also effects of diet on cortisol levels and body weight. However, it should be pointed out that the *ad lib* hay group had higher bodyweights (as they were not accustomed to such a diet) than the other groups.

The final study in this series evaluated the effects of feeding forage, without any grain, on substrate utilisation and performance during exercise. In this instance the diets were:

1. *Fast*: overnight fast; Grain and hay was fed during the adaptation phase but not on the day of the CET.
2. *Ad lib hay*: For 7 days before the CET the horses were offered 5- 6kg grass hay at 2200h and 1.13kg at 1200 and ~2.4kg at 1600. If all the hay ad been consumed before the CET they were offered 2kg hay that morning. Grain was fed during the adaptation phase but not on the day of the CET.
3. *Am hay*: ~2.3kg hay 2200hrs on the day before and ~2.3kg 3 hrs before the CET. Grain was fed during the adaptation phase but not on the day of the CET.
4. *Grazing*: for 7 nights before the CET they were housed in small grass paddocks overnight and then kept in stalls and offered grain and hay. On the morning of the CET the horses were removed from the paddocks at 0600 and no hay was given.

There was relatively little effect of these different feeding practices on the glucose response although it tended to be lower in the Am hay group for most of the exercise see Figure 2.

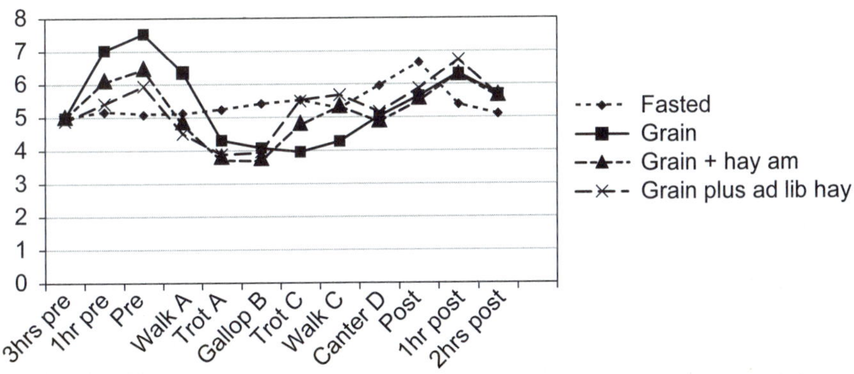

Figure 1. Glucose concentrations (mmol/l) before during and after a simulated competition exercise test.

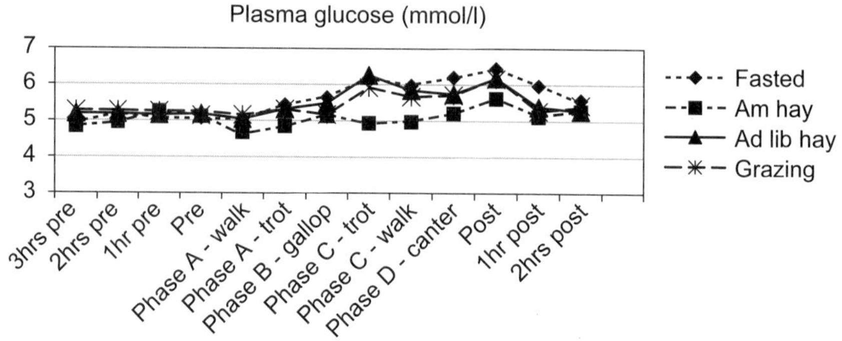

Figure 2. Plasma glucose concentrations (mmol/l) before during and after a simultated competition exercise test.

Hay feeding did not affect FFA nor cortisol levels. In this study there was no significant effect of the *ad lib* hay on total protein concentrations perhaps because they had adjusted to the *ad lib* feeding regimen. Feeding forage before exercise therefore did not appear to adversely affect performance (other than the potential to increase body weight and possibly reduce plasma volume if the intake is not controlled). The horses that had been grazing before the CET tended to be heavier but did not show the same apparent changes in plasma volume or heart rate as they did with the unaccustomed *ad lib* hay feeding in the previous study, perhaps because they were better able to equilibrate fluid between the gut and the plasma.

In a study looking at more intensive exercise (2 mins at 15% VO_{2max}) the effect of restricted hay intake (for 3 days at 1% BW ~ 4.5kg hay) compared to *ad libitum* feeding (~10kg hay) was evaluated (Rice *et al.*, 2001). Feed and water were removed 4hrs before the exercise. The restricted hay intake resulted in a decrease of approximately 2% in BW (528 vs. 539kg), which was associated with an increase in the mass specific rate of oxygen consumption during sprint exercise (243 vs. 233ml/kg/2min) with a corresponding decrease in anaerobic energy expenditure. Peak lactate was higher in the adlib fed horses (22.2 vs. 19.1 mM). This suggested that it might be beneficial to reduce hay intake to the recommended minimum of 1%Bwt/day. However, as very few horses in intense performance work are currently fed *ad lib* forage, further work is needed to determine the exact nature and level of optimal amount for horses in intensive work.

Conclusions

These studies confirm that insulin is a potent inhibitor of lipolysis and fatty acid oxidation in skeletal muscle, and also promotes glucose uptake into muscle (via recruitment of the transporter protein GLUT4 to the sarcolemma). Thus, hyperinsulinemia at exercise onset will tend to suppress non-esterified fatty acid (NEFA) availability plus lipid oxidation and increase reliance on carbohydrate stores (including plasma glucose) for energy production. This decrease in plasma glucose concentration may be short-lived so that during prolonged moderate-intensity exercise (e.g. 60 min at 50% of VO_{2max}), the plasma glucose concentrations of grain fed horses are not substantially different from horses fasted before exercise. On the other hand, plasma NEFA and lipid oxidation remain lower when compared to the fasted state throughout exercise (Jose-Cunilleras *et al.*, 2002). It is therefore currently recommended that for horses competing in endurance, eventing and show jumping they should not be given cereal based meals within 2 - 3hrs of the competition. Large meals (hay or grain or a combination) consumed near the start of exercise may result in a decrease in plasma volume as a result of fluid shifts into the gastrointestinal tract. Such reductions in plasma volume could compromise cardiovascular and thermoregulatory function during exercise. Total feed restriction, however, prior to competition is not recommended as it increases the risk of gastric ulceration.

Fortunately as suggested above small forage meals consumed 2 to 3 hours before exercise may have minimal effect on substrate availability and oxidation during sustained exercise. It is commonly now recommended that horses are allowed to nibble on small forage meals (1-2 kg, as fed) in the 1-3 h period pre-endurance/3DE/show jumping.

This series of studies therefore highlights the role that research can play in defining feeding recommendations that both support performance and the welfare of the competition horse.

Effect of adding chaff

It is not always possible to work horses, especially during training, 3hrs after a small meal. However, especially in quick or greedy eaters eating a concentrate-based meal can result in a significant decrease in plasma volume. Adult horses, around 500kg BW, may secrete more than 100L of fluid per day into the pre-caecal section of the gastrointestinal tract i.e. around 70-100ml/minute (Harris *et al.*, 2006). When large meals of either pellets or cereal grains are fed infrequently a transient state of hypovolaemia occurs as a result of meal stimulated upper gastrointestinal secretions. Within one hour of feeding ponies there has been shown to be on average a 15% loss in plasma volume. In quick or greedy feeders this can be up to a 24% loss (Clarke *et al.*, 1990). Slow feeders compensate whilst they eat and the resultant volume losses are therefore small. These changes in plasma volume are not seen when frequent small meals are fed.

Slowing down quick and greedy feeders can obviously be advantageous. There are a number of possible ways to do this. In particular we have been evaluating the role of including supplemental chaff at an appropriate chop length. In one study in Australia, the effects of including chaff at the time of feeding, were evaluated in six thoroughbred geldings fed a meal of oats (3 g/kg bodyweight) each morning (Campbell *et al.*, 2005). Longer or shorter ground lucerne chaff (4 cm and <2 cm, respectively) was included with this meal at five inclusion rates ranging from 7% to 37.5% of the total mixed ration. As the amount of chaff was increased the rate of feed intake declined with maximal effect when chaff made up 33% of the ration. The rate of food intake was unaffected by chaff length. This was also found in another series of investigations carried out in Australia. This time they evaluated the effect of including short (<2 cm) lucerne chaff with either a sweet concentrate mix (cereals, pellets, coating: Harris *et al.*, 2005a) or an oat meal fed at 300g/100kg BW (Harris *et al.*, 2005b) at 6 and 35% addition on top of the core ration. The rate of intake (over 3 consecutive days) as well as the glycaemic response to a 2kg meal of hay was initially determined in 8 light-horse geldings before the oats or sweet feed was fed. The rate of feed intake was slowest during the period that hay was fed in both studies (31g/min). The sweet feed and oats were ingested at about twice the rate of hay (77 and 58g/min respectively). The addition of 6% chaff to the concentrate mix did not alter the rate of intake, but an apparent decrease in the rate of intake was seen with the 26% chaff (49 and 52g/min for the sweet feed and oats respectively) although this was not statistically significant. This was due in part to an increased volume of intake per minute, particularly with the oat meal see Figure 3. Under certain circumstances this increase in the intake volume, if associated with reduced chewing, could increase the risk of complications such as choke.

There was no correlation between the area under the curve (AUC) for blood glucose concentrations and rate of feed intake. The AUC and peak blood glucose level was significantly higher for the sweet feed compared with hay, and there was no significant effect of chaff inclusion on these parameters for either the oats or the sweet feed. This suggests that there is no significant adverse effect on starch digestion in the small intestine, as evaluated by the glucose response, of such a chaff addition.

Another study which looked at the effect of including 10%, 20% or 30% of straw chopped into either 2.5cm or 4cm lengths to a trial pellet fibre mix (containing 10% short cropped lucerne: Ellis *et al.*, 2005) also showed no effect of chaff length. Around 1 kg of the mix was fed to horses each morning

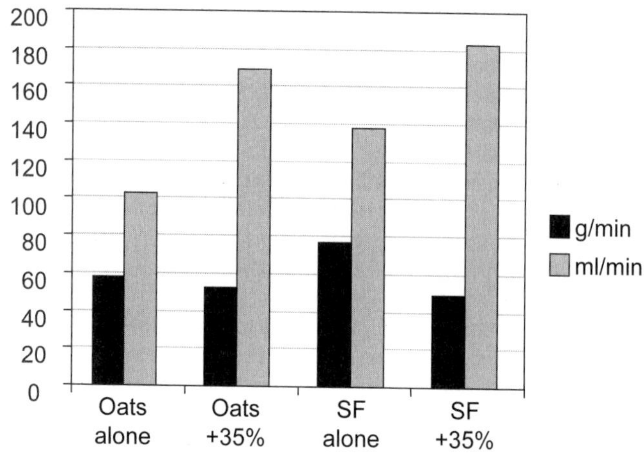

Figure 3. Amount (g/min) and volume (ml/min) of oats and sweet feed fed alone or with a 35% addition of chaff.

and evening and the time taken for horses to eat the various feeds as well as the average number of chews per minute were monitored. The rate of chewing remained about the same regardless of the type of food fed. However, the type of food had a significant effect on the rate of intake and the total number of chews per kg. Diets containing added straw were eaten much more slowly than their normal diet mix, (containing 50% pellets, 15% lucerne-straw chaff, and 35% soaked sugar beet pulp), their normal pellets and the trial pellets on their own. There was, however, no significant effect of adding 10% straw chaff to the trial pellets with 10% choped alfalfa but there was a significant effect of the 20% and 30% addition. Whilst adding more than 20% of the straw chaff had no significant effect the intake rate did decrease and the time taken to eat the meal increased. Average food intake times ranged from 7 min/kg on the normal diet to 16 min/kg when the longer chaff was included at 30% to the pellet/chaff mix.

Conclusion

These studies suggest that including chaff at levels of around 25 - 30% inclusion rate can lead to horses spending more time eating their meal which, in turn, is likely to lead to increased saliva production and potentially improved buffering of the gastric contents (Harris *et al.*, 2006). These studies did not show the same apparent adverse effect on glycaemic response that was seen with the larger amounts of hay (Pagan and Harris 1999) which may represent a difference between short and long chop fibre. Again this work has led to the recommendations of a feeding practice that is of benefit of the horse but does not appear to adversely affect performance. Further work is needed, however, to evaluate optimal lengths and types of chaff.

Feed intake behaviour

Interestingly while there have perhaps been great advances in the last 15years in our understanding of the dietary requirements of horses, as well as the functionality of nutrients, the way in which horses eat and are fed in domestic environments has attracted comparatively little research interest. For humans and domestic pets, eating involves more than simply the ingestion of food. It encompasses aspects such as sensory stimulation and social interaction. Proper selection and mastication off food is also essential for the maintenance of digestive health.

Side preferences

In nutritional studies, food preferences are often determined by measuring the relative intakes of two food choices positioned beside another. One possible source of error in such studies is the consistent selection of one food based on its position, i.e. left or right. This is termed side preference feeding and was investigated by recording food intake and feeding behaviours when the same feed was offered to horses in two separate containers (Bottom *et al.*, 2005). Eight of the 18 horses in the study (44%) demonstrated the development of significant side preferences over time, with an equal number of horse showing preferences for food positioned on the left or right. Interestingly, when the horses were divided into age groups, the incidence of side preference was higher in older (16-22 years of age) horses than younger (2-4 years) and middle aged (8-14 years) horses. The strength of the side preference was also greater in the older horses. The time that horses spent eating also increased with age, whereas the young horses tended to spend more time exploring the feed and the environment rather than ingesting food (Table 1).

Reduction in exploratory behaviour with age may explain the observed rise in the incidence and strength of side preferences, since these horses rapidly and exclusively choose to consume one of the feed offerings. These findings have important implications for the design and interpretation of two-choice preference tests, particularly when they are used in middle and old age horses.

Table 1. Percentage of time spent in locomotory or non-ingestive behaviour during a meal.

	Non-ingestive (%)	Locomotory (%)
Young	28.4	7.6
Middle	4.7	0.5
Old	0.1	0.0

Flavour choice

Like most grazing herbivores, horses are thought to select their food according to its sensory properties and specifically visual cues, odour, taste, texture, availability and variety. There is, however, little published information on the role of flavour in food selection by domestic horses. To address this, the effects of flavour and concentrate formulation on food selection and acceptance were investigated in a series of trials. The acceptance/rejection and quantity and rate of consumption of cereal by-product meals containing 15 separate flavours was assessed. Twelve flavours were universally accepted by the eight horses tested, demonstrating that horses are attracted to a wide range of flavours (Goodwin *et al.*, 2005a). The eight most readily consumed were subsequently presented in paired preference tests. From these tests, the flavours were ranked in descending order of preference as fenugreek, banana, cherry, rosemary, cumin, carrot, peppermint and oregano. In the final part of this study, the consumption of mineral pellets flavoured with fenugreek or banana was found to be greater than that of unflavoured pellets (mean 66 vs. 52 vs. 195 seconds for the Fenugreek, banana and unflavoured pellets respectively). There was no difference between the fenugeek and banana flavoured pellets. These findings demonstrate that, in the short-term at least, flavour does have a significant impact on diet selection, acceptance and consumption and may help to ensure that horse ingest certain feedstuffs that some may try and avoid.

The effects of sensory variety on the selection and ingestive behaviour of horses was also investigated in a separate set of trials (Goodwin *et al.*, 2005b), which looked at the influences of concentrate formulation, physical composition and flavour. The flavours assessed were molasses, mint, herbs and garlic. All of the flavoured concentrates were well accepted by the eight horses in the study but significant changes in concentrate preference were observed between and within trials. It was observed that horses repeatedly interrupted the consumption of a concentrate, for which they had previously demonstrated a preference, to forage on other diets that were presented in 'multiple taste tests'. This meant that their consumption of the preferred concentrate was always lower in multiple compared with single test sessions. The total time spent foraging was, however, higher in these multiple sessions. The basis for switches in preference is likely to reflect an evolutionary adaptation that allows horses to explore and exploit the heterogeneity of dietary resources available. This would allow them to choose a diet of better quality than the average vegetation on offer. These trials showed that manipulating a single sensory characteristic of a diet is sufficient to affect both foraging and non-foraging behaviours, for example horses stood inactive for longer periods when only a single concentrate was available.

Environmental enrichment

Concentrate feeds are usually offered to horses in buckets or mangers and this means of presentation does nothing to slow the rate at which these foods are consumed. Several so-called foraging devices are available commercially and can be used to provide part of the ration in a less easily accessible manner and provide sensory stimulation. Some of these operant foraging devices are based on the Edinburgh foodball (Winskill *et al.*, 1996) which aim to prolong food-handling time and are designed to be presented on the floor. The ration is placed inside these devices and horses must either nuzzle the unit or move it their head or a foot to release food. The one major drawback to these current devices is the fact that they are placed on the floor, raising concerns about the ingestion of foreign materials with the food. An alternative is to use foraging devices that fit inside feed mangers or buckets and the behavioural benefits of these were therefore evaluated (Goodwin *et al.*, 2007). Horses were presented with three devices varying in sensory complexity - round, square and polyhedral in design - in either a manger or bucket and their behaviour recorded. All horses foraged successfully from at least one of the devices but also exhibited some frustration, either biting or pawing at the devices, and frequently removed the devices from the buckets. This was not the case when the devices were placed in feed mangers. The polyhedral design appeared the most promising in that it was consistently and repeatedly associated with the longest duration of foraging. However, the study did suggest that when the dispensing of pellets, as a means of enriching the environment, occurs in an unpredictable way, the unpredictability of reward could be a source of frustration.

Benefits of a varied diet

Horses tend not to eat a discrete meal from one plant nor do they concentrate solely upon one species of plant even if it is a preferred plant species (Mariner, 1980). Horses appear to have evolved by consuming a varied and *ad libitum* roughage diet typically high in fibre and this contrasts with the restricted forage diet that most stabled horses are fed. The feeding behaviour of free ranging horses may be more complex than those kept in stables (Carson and Wood-Gush, 1983) as the diet of free ranging horses tends to be very varied on a daily basis and from season to season (Gill, 1988; Hansen, 1976). Previous studies have linked this to changes in patterns of behaviour, including a tendency for horses to eat their straw bed, a habit that can lead to colic.

In order to start to investigate this the behaviour of 12 competition horses was monitored during a number of trials where they were either provided with a single food or six types of feed (Goodwin *et al.*, 2002). The work showed that, at least in the short term, by supplying multiple 'forages' it was possible to significantly affect the behaviour of the stabled horses (promoting more natural foraging behaviour patterns) whilst reducing the consumption of straw as well as the amount of non- foraging behaviours such as 'weaving' and 'pawing'. This has been followed up by more long term studies in 9 horses, aged 5-20 years and of mixed breeding, which were fed either a single forage (hay) or multiple forages (three short-chop and three long-chop commercially available forages) in two daily feeds for 9 days, and then fed the alternate ration (Thorne *et al.*, 2005). The short forages were presented in individual buckets, whereas the long forages were provided in separate hay nets. The horses were observed on days 3, 5, 7 and 9 of each period for the 25 minutes following forage delivery in the morning and afternoon. This study showed that when fed the multiple forages, horses performed foraging behaviour significantly more frequently and for longer periods than when fed the single forage. Foraging on the six sources offered differed in frequency, latency and duration, and, while clear preferences were demonstrated, all forages were sampled during the periods of observation. Horses on the single forage spent significantly longer periods performing behaviours thought to represent a search for alternative food sources, such as moving and looking out of the stable. This type of enrichment may offer horse owners a practical and easily adopted means of facilitating foraging behaviour in their horses, and may help prevent the occurrence of stereotypical and redirected behaviours.

Is the presence of another horsed beneficial or stressful during feeding?

Increasing visual contact between horses in a stable setting is often anecdotally assumed to be beneficial to the well being of horses. There has, however, been limited research into the relationship between neighbouring horses, especially at potentially stressful periods such as feeding. Whether the presence of a familiar horse during meal times is of benefit or not was therefore investigated in Scotland (Redgate *et al.*, 2004). Six horses were fed alone in their own stable and then in a familiar arena in the presence of another horse that was eating or not eating. The horses were fed a standard plain concentrate feed and behaviour was recorded every 5 seconds for a 5-minute period. There were individual differences in the amount of vigilant behaviour recorded, which was expressed as 'eat hover' (head hovering over the bucket whilst chewing, head below the withers) and 'eat head up' (chewing with head above withers). These feeding behaviours were thought to allow the horse to survey the environment. Some horses showed more vigilant behaviour in the presence of another horse than when stabled alone, and this was regardless of whether the other horse was eating or not. Conversely other horses showed more vigilant behaviour when fed in the stable alone. Further work is required to establish if this effect is seen with larger groups of horses and to identify stable designs that best cater for individual, social and behavioural needs.

No of meals

Another potential method for enriching the stable environment and reducing stereotypic behaviours is to provide more frequent, but smaller, concentrate meals. This was evaluated by splitting the daily concentrate ration into two, four or six equal feeds (Cooper *et al.*, 2005). Prior to the study, baseline behavioural characteristics were recorded and, while these showed a low incidence of stereotypic behaviours, it was clear that such behaviour was more commonly seen in the afternoon which coincided with there being less forage available. The incidence of oral stereoptypies - defined as repetitive oral activity without ingestive intention, such as sham chewing, tongue rolling, and biting, chewing and licking of stable fittings - decreased as the number of meals increased. There was, however, a concomitant increase in anticipatory-related activities like weaving and nodding prior to feeding. The incidence of weaving, nodding and oral stereotypic behaviours also increased in the control horses that continued to be fed twice a day.

As suggested by these authors increasing mealtimes may lead to an increase in post-feeding stereotypy, unless for example the individual meal-sizes are large enough to be nutritionally satisfactory without being so large as to cause digestive dysfunction or the horses have access to an alternative more acceptable means of expressing feeding behaviour such as access to forage (Goodwin *et al.*, 2002).

Conclusions

These studies suggest for example that providing sensory variety may present a means of increasing dietary variety without drastically changing the nutrient content of the ration, and this is likely to be of particular benefit for horses fed restricted or monotonous diets. There also appear to be behavioural advantages from providing multiple forages to stabled horses by enriching their environment through offering variety and enabling the expression of foraging behaviours that more closely mimic the patch foraging seen in free-roaming horses.

These results also indicate that the individual nature of horses needs to be considered when determining their optimal management. There are, for example, differences between horses in respect of their comfort in the presence of another horse during meal times, with some finding it apparently less stressful and others more stressful than being fed alone.

Finally these studies highlight the complexities of trying to modify equine behaviour in a yard situation, where there are multiple stables and interactions between neighbouring and distant horses. While increasing meal frequency did reduce the incidence of some stereotypies, it actually increased these behaviours in horses that were in visual contact and not fed at the same time.

Diet and behaviour

There continues to be considerable interest in the possible links between diet and behaviour in humans and various animal species. Horses have not escaped this attention, especially as the process of domestication has necessitated changes in their freedom and diet that potentially impact upon their behaviour. For instance, the development of stereotypic behaviours like weaving and crib-biting in stabled horses has been linked with certain feeding practices, most notably high concentrate and/or low forage diets (McGreevy *et al.*, 1995; Waters *et al.*, 2002). Many people, in developed countries, have horses in order to ride them or to watch them being ridden. Being naturally fear and flight animals, their behaviour becomes increasingly important when being ridden in environments beyond the rider's control - as is often the case!

Currently much of the interest around diet and nutrition is focused on the areas of temperament and tractability. In the late 1990's we reviewed a number of studies that we had been carrying out in this area in America with Colleagues in Virginia (Kronfeld *et al.*, 1999). One of these studies looked at spontaneous locomotion as measured by pedometers and also evaluated the 'startle response' to an unusual object (abruptly opening umbrella). The study showed that the replacement of cereals with oil and fibre might diminish spontaneous locomotion and reactivity. Other work confirmed objectively that horses can show differences in behaviour when fed cereal based diets compared with high forage diets (Davidson *et al.*, 1998). It has also been shown that the addition of oil to the diet may have a beneficial effect on behaviour (Holland *et al.*, 1996a). However, the precise mechanism for these dietary effects on behaviour have not been elucidated to date. The paper goes on to describe some of our more recent work in this area.

Diet and the stress of weaning

One of the most stressful times in a horse's life is thought to be weaning. Back in 1996 studies that compared the effects of diet and weaning method on the stress responses in thoroughbred foals raised on pasture were published (Holland *et al.*, 1996b). The work confirmed that foals weaned abruptly appeared to be more stressed than those weaned more gradually. However, perhaps more interesting were the findings that the cortisol levels were higher in foals fed a more traditional starch and sugar based diet than those fed the alternative fat and fibre rich concentrate. This was reflected in the foals fed the fat and fibre rich diet, appearing to be less stressed, suggesting that diet may affect a foal's ability to handle the stress of weaning.

A later study compared the effects of diet and the use of probiotics on weaning stress (Swanson *et al.*, 2003). Dietary supplements, including probiotics, are often given to horses to help maintain digestive function and to help reduce other stressors, such as travelling, and might therefore be beneficial during weaning. In this study half the foals were fed a commercial probiotic (Probios®) containing lactic acid producing bacteria and the remainder received a placebo from 4 days prior to weaning. Plasma and faecal samples were collected from 4 days prior to, and for 4 days after, weaning and the behaviour of the foals was analysed for measures of stress. The degree of stress experienced by the foals was also assessed by adrenocorticotrophin hormone response tests conducted 48 hours after weaning. The foals fed the fat and fibre concentrate were found to have significantly higher plasma concentrations of IgA, IgG and α-tocopherol, and had higher faecal pH than those fed the sugar/starch supplement (Figure 4).

Foals that received the probiotic had higher faecal lactate and lower faecal acetate concentrations, suggesting that the lactic acid bacteria had survived transit to the hindgut and established colonies there. There was not, however, any effect of the probiotic on plasma markers of immune status or faecal measures of bacterial activity, and no effect of either diet or probiotic on measures of stress. These results indicate that supplementation with a concentrate rich in fat and fibre may improve immune status and encourage a more diverse microbial population in the hindgut, thus ensuring better adaptation to the post-weaning diet, perhaps helping foals to better withstand the physiological stress of weaning.

Another study (Ordakowsk *et al.*, 2003) looked primarily at foal feed intake behaviour but it was noted that foals with access to the fat and fibre pasture supplement (FFS) appeared to be more confident in their environment than those fed the starch and sugar rich variant (SSS). The behavioural aspects were explored in more detail in a later publication (Redgate and Davidson, 2007) where it was confirmed that eight out of the ten FFS weanlings appeared to be more relaxed in the test environment in contrast to four out of the ten for the SSS weanlings (Fishers Exact Test, $p=0.08$ (Figure 5), based on locomotory patterns, interaction with the food provided and vocalisation.

These observations were explored in more detail in the second study, where behaviour was observed in 17 foals from the age of 2 to 40 weeks. Each foal received either a starch and sugar (SS) diet

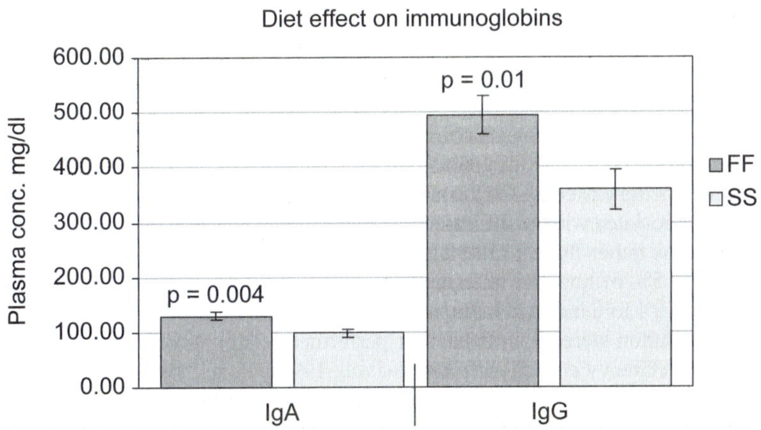

Figure 4. Effect of feeding a supplement rich in Fat and Fibre (dark grey) or starch and sugar (light grey) on plasma immunoglobulin concentrations around weaning.

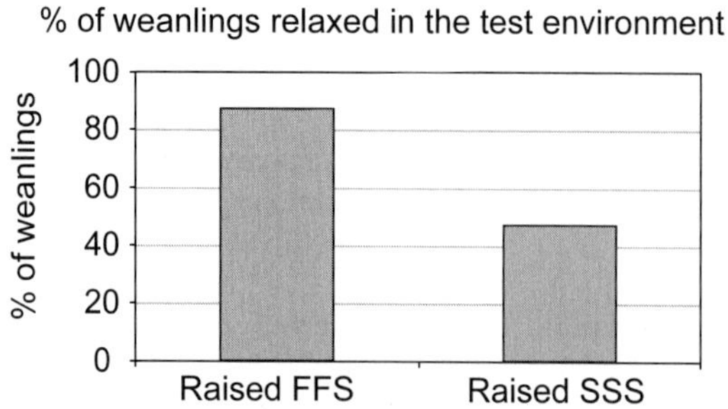

Figure 5. Percentage of weanlings that were relaxed in the text environment fed on starch and sugar rich supplement (SS) or fat and fibre rich supplement (FF).

or a fat and fibre (FF) diet. The two diets contained similar digestible energy, crude protein and micronutrients, but differed in the fat and non-structural carbohydrate content. Aspects of behaviour were observed every 2 weeks from 2 to 40 weeks of age (Nicol *et al.*, 2005). Approximately 2 months after weaning, the temperament and tractability of the young horses were assessed using standardised tests. Responses to a novel object, to a novel person, and during a handling test were determined. Foals grew well on both diets and there was no effects of diet on behaviour prior to weaning. Immediately after weaning, foals receiving the fat and fibre supplement cantered less frequently, for a shorter duration, and appeared to be more settled than those fed the starch and sugar supplement. The fat and fibre foals also spent significantly more time investigating and less time looking at a novel object, spent less time walking away from a novel person and completed a handling test in a quicker time.

Overall, the authors concluded that 'the horses that received the FF diet appeared less distressed immediately after weaning and seemed calmer and more inquisitive during a range of temperament tests'.

Conclusions

These studies are thought to be the first research to objectively show that diet can influence behaviour, reactivity and stress responses in normal horses.

Diet and abnormal behaviour

Certain specific behaviour patterns, once thought to be due to boredom or naughtiness, may in fact be linked to modern management practices, and specifically diet in many cases (although stable confinement is also a major factor). Stereotypies have been described as invariant and repetitive behaviour patterns that seemingly have no function and are fairly commonly seen in the domesticated horse. In a 4 year prospective study of young thoroughbred and part thoroughbred horses stereotypic and redirected behaviours affected ~35% of the animals studied (Waters *et al.*, 2002). Weaning by confinement was associated with an increased rate of development compared with paddock weaning and housing in barns rather than at grass after weaning resulted in a further increase. Crib-biting was initiated by 10.5% of horses with a median age of 20 weeks and feeding concentrates after weaning resulted in a 4 fold increase in the rate of development of crib-biting (Waters *et al.*, 2002). Crib-biting is a common stereotypic behaviour performed by approximately 5% of adult stabled domestic horses (McGreevy *et al.*, 1995; Redbo *et al.*, 1998; Nicol, 1999) although interestingly it has not been recorded in wild equids. Although the cause(s) of crib-biting are not known, risk factors recently identified include the feeding of high concentrate and, or, low forage diets as well as stable design etc., (McGreevy *et al.*, 1995; Waters *et al.* 2002). Horses only produce saliva when they chew and recent work investigated the hypothesis that young horses that start to crib bite have gastric mucosa abnormalities (Nicol *et al.*, 2002). It was postulated that crib biting provided some relief by helping to produce alkaline saliva which would in turn help buffer the stomach pH especially when other opportunities for mastication were limited. The aim of this work was therefore to evaluate the link between the condition of the stomach and the incidence of cribbing in young foals before the condition becomes well established (when the initiating cause may no longer be present).

Nineteen young horses (weanlings) were observed that had recently started to perform crib-biting, together with 16 non-stereotypic similarly aged horses. The project included the gastroscopic examination of 15 crib-biting and nine normal foals, who were then randomly assigned to either a control or antacid diet (the same base complementary ration with the addition of Neighlox: KPL Technology Inc.) for 3 months and the stomachs re-examined. Behaviour was monitored by direct observation at various points throughout the 3-month period. At the start of the study: within the crib-biting foals 20% (3) had ulcers only, 13% (2) had inflammation only, 27% (4) had both ulcers and inflammation, and 40% (6) had no sign of ulcers or inflammation. Of the normal foals, 11% (1) had ulcers only, none had inflammation only, 11% (1) had both ulcers and inflammation, and 78% (7) had no sign of ulcers or inflammation. The ulcers were few in number and generally mild with the exception of 3 of the crib-biting foals that had moderate, heavy or severe ulceration. Crib-biting foals with ulcers at the start of the trial had been performing stereotypic behaviour for a mean of 83 days, whilst foals without ulcers at the start of the trial had been performing stereotypic behaviour for a mean of 114 days. Overall at the first gastroscopy, the stomachs of crib-biting foals were found to be significantly more inflamed and ulcerated than those of the normal foals. There is, however, likely to be a continuum in stomach condition from normal foals, through mild crib-biters to more severe crib-biters. The antacid diet resulted in a significant improvement in stomach condition overall ($p<0.05$) - although in those with marked ulceration total resolution did not occur. Crib- biting behaviour declined in most foals, regardless of diet but tended to show a greater decline in foals on the antacid diet ($p<0.1$) especially where there had been an improvement in stomach condition. Foals that showed the greatest reduction in ulceration score also tended to show the greatest reductions in crib-biting frequency and duration.

In an additional pilot study the authors placed six foals on a specific weanling diet where the energy came predominantly from fat and fibre rather than from more traditional sugar and starch sources. This diet was designed in part to increase the time spent chewing and two foals on this diet stopped crib-biting after 3 months and one foal significantly reduced the incidence of crib-biting.

Conclusions

These novel results suggested an association between stomach condition and abnormal behaviour that can be modulated through diet. These results add weight to the belief that crib-biting is initiated in foals in response to more acid stomach conditions, or to increased sensitivity to normal levels of acidity. Altering gastric acidity may, therefore, modify the expression of this behaviour and this was supported by the changes in gastric ulceration observed.

Summary

The above provides a brief insight into some of the work that has recently been undertaken to explore the influence that the way we feed and manage our horses can have on their behaviour, welfare and performance. The end result may not only reflect the practices used but also the individual horse's response to such practices. This individuality is an important consideration when we look at means to improve the way that we manage and keep the animals in our care.

References

Bottom, S.H., R.E. Lawson, H. Owen, S. Hall and P. Harris, 2005. Equine feeding side preference; implications for two-choice preference testing. In, Proceedings of the 19th Equine Science Society Symposium, Tuscon:47-48.

Campbell, T.E., P.A. Harris, H.C. Doughty and M.N. Sillence, 2005. Effect of chaff quantity and length on rate of intake in horses fed a concentrate diet. Asia Pacific Journal of Clinical Nutrition; 14 (Suppl.):S80.

Carson, K. and D.G.A.M. Wood-Gush, 1983. Equine Behaviour 11 Review of the literature on feeding, eliminative and resting behaviour Appl Animal Ethol 10:179 -190.

Clarke, L.L., M.C. Roberts and R.A. Argenzio, 1990. Feeding and digestive problems in horses: Physiologic responses to a concentrated meal. Vet. Clin. North Am. Equine Pract. 6: 433-50.

Cooper, J.J., N. Mcall, S. Johnson and H.P.B. Davidson, 2005. The short-term effects of increasing meal frequency on stereotypic behaviour of stabled horses. Applied Animal Behaviour Science 90:351-364.

Davidson, N. and P. Harris, 2002. Nutrition and Welfare, In: The welfare of horses, N. Waran (Ed.). Kluwer Academic publishers Netherlands 45-76.

Davidson, N., J. Pagan, D. Goodwin, P.A. Harris and S. Cook, 1998. The Effects of Diet and Exercise on the Behaviour of Stabled Horses. 32nd Congress of ISAE.

Duren, S.E., J.D. Pagan, P.A. Harris and K.G. Crandell, 1999. Time of feeding and fat supplementation affect plasma concentration of insulin and metabolites during exercise. Equine Vet J Suppl 30:479-484.

Ellis, A.D., S. Thomas, K. Arkell and P. Harris, 2005. Adding chopped straw to concentrate feed: The effect of inclusion rate and particle length on intake behaviour of horses. Pferdeheikunde; 21:35-36.

Gill, E.L., 1988. Factors affecting body condition of New Forest ponies PhD thesis Department of biology University of Southampton.

Goodwin, D., H.P.B. Davidson and P. Harris, 2002. Foraging enrichment for stabled horses: effects on behaviour and selection, Equine Vet. J., 34 (7):686-691.

Goodwin, D., H.P.B. Davidson and P. Harris, 2005a. Selection and acceptance of flavours in concentrate diets for stabled horses. Applied Animal Behaviour Science 95:223-232.

Goodwin, D., H.P.B. Davidson and P. Harris, 2005b. Sensory varieties in concentrate diets for stabled horses: effects on behaviour and selection. Applied Animal Behaviour Science 90:337-349.

Goodwin D., H.P.B. Davidson and P.A. Harris, 2007. A note on behaviour of stabled horses with foraging devices in mangers and buckets. Applied Animal Behaviour Science 105:238-243.

Hansen, R.M., 1976. Foods of free roaming horses in Southern New Mexico. J Range Management 29(4):437.

Harris, P.A., 1997. Feeds and Feeding in the United Kingdom. Robinson, N.E. (Ed.) Current Therapy In Equine Medicine 4: WB Saunders: 698-703.

Harris, P.A., 2000. Feeding Practices in the UK and Germany. In KER proceedings 241-267.

Harris, P.A., M. Sillence, R. Inglis, C. Siever-Kelly, M. Friend, K. Munn and H. Davidson, 2005a. Effect of short (<2cm) lucerne chaff on the intake rate and glycaemic response of sweet feed. Pferdeheikunde 21:87-88.

Harris, P., M. Sillence, R. Inglis, C. Siever-Kelly, M. Friend, K. Munn and H. Davidson, 2005b. Effect of short Lucerne chaff on the rate of intake and glycaemic response to oat meal. In, Proceedings of the 19th Equine Science Society Symposium, Tuscon:151-152.

Harris, P., M. Coenen, D.L. Frape, L.B. Jeffcott and H. Meyer, 2006. Equine nutrition and metabolic disease. In Equine Manual Higgins A and SnyderJ (eds) Elsevier Saunders London p 151-222.

Holland, J.L., D.S. Kronfeld and T.N. Meacham, 1996a. Behavior of horses is affected by soy lecithin and corn oil in the diet. J. Anim. Sci. 74:1252-1255.

Holland, J., D.S. Kronfeld, R. Hoffman, K.M. Greiwe-Crandell, T. Boyd, W.L. Cooper and P. Harris, 1996b. Weanling stress is affected by nutrition and weanling methods. 2nd European Conference On Horse Nutrition: Nutrition And Nutritional Related Disorders Of The Foal. 257-260.

Jose-Cunilleras, E., K.W. Hinchcliff, R.A. Sams and S.T. Devor, 2002. Glycemic index of a meal fed before exercise alters substrate use and glucose flux in exercising horses. J. appl. Physiol. 92:117-128.

Kronfeld, D.S., S.E. Custalow, P.L. Ferrante, L.E. Taylor, J.A. Wilson and W. Tiegs, 1998. Acid -base responses of fat adapted horses: relevance to hard work in the heat. Applied animal behaviour science 59:61-72.

Kronfeld, D.S., J.L. Holland, R.M. Hoffman and P.A. Harris, 1999. Dietary influences on behaviour and stress. Equine Veterinary Journal Suppl 28- The Role of the Horse in Europe: Proceedings of the Waltham Symposium: 64.

Mariner, S., 1980. Selective grazing behaviour in horses PhD thesis Univ of Natal Durban s Africa.

McGreevy, P., C.J. Nicol, P. Cripps, L. Green and N. French, 1995. Management factors associated with stereotypic and redirected behaviour in the thoroughbred horse. Equine vet. J. 27:86-91.

Nicol, C.J., H.P.D. Davidson, P.A. Harris, A.J. Waters and A.D. Wilson, 2002. Study of Crib-biting and gastric inflammation and ulceration in young horses. Veterinary Record 30:658-662.

Nicol, C.J., A.J. Badnell-Waters, R. Rice, A. Kelland, A.D. Wilson and P.A. Harris, 2005. The effects of diet and weaning method on the behaviour of young horses. Applied Animal Behaviour Science 95:205-221.

Nicol, C.J., 1999. Understanding equine stereotypies. Equine vet. J. Suppl. 28:20-25.

Ordakowski, A.L., H.P.B. Davidson, S.E. Redgate, P.A. Harris and D.S. Kronfeld, 2003. Characteristics of foal feeding behaviour. WALTHAM International Symposium, Bangkok 2003; 30.

Pagan, J.D., I.H. Burger and S.G. Jackson, 1995. The influence of time of feeding on exercise responses in thoroughbreds fed a fat supplemented or high carbohydrate diet. Equine Nutrition and Physiology 14th symposium proceedings p92-93.

Pagan, J.D. and P.A. Harris, 1999. The effects of timing and amount of forage and grain on exercise response in Thoroughbred Horses. Equine Vet J. Suppl 30 Jeffcott L (ed): 451-458.

Redbo, L., P. Redbo-Torstensson, F.O. Odberg, A. Hedendahl and J. Holm, 1998. Factors affecting behavioural disturbances in race-horses. Anim. Sci. 66:475-481.

Redgate, S.E. and H.P.B. Davidson, 2007. The effect on behaviour of the presence or absence of a familiar horse during feeding. In: Horse behaviour and welfare, M. Hausberger, E. Søndergaard and W. Martin-Rosset (Eds.), Wageningen Academic Publishers, the Netherlands. p.65-69.

Redgate, S.E., A.L. Ordakowski-Burk, H.P.B. Davidson, P.A. Harris and D.S. Kronfeld, 2004. A preliminary study to investigate the effect of diet on the behaviour of weanling horses. In, Proceedings of the 38th International Congress of the ISAE, Helsinki, p154.

Rice, O., R. Geor, P. Harris, K. Hoekstra, S. Gardner and P. Pagan, 2001. Effects of restricted hay intake on body weoght and metabolic responses to High intensity exercise in thoroughbred horses. Proceedings of Equine Nutrition and Physiology society Lexington 273-279.

Swanson, C.A., R.M. Hoffman, D.S. Kronfeld and P.A. Harris, 2003. Effects of diet and probiotic supplementation on stress during weaning in Thoroughbred foals. Proc. 18th Equine Nutrition and Physiology Symposium 2003; 243.

Thorne, J.B., D. Goodwin, M.J. Kennedy, H.P.B. Davidson and P. Harris, 2005. Foraging enrichment for individually housed horses: Practicality and effects on behaviour. Applied Animal Behaviour Science 94:149-164.

Waters, A.J., C.J. Nicol and N.P. French, 2002. Factors influencing the development of stereotypic and redirected behaviours in young horses: the findings of a four year prospective epidemiological study. Equine vet. J. 34:572 -579.

Winskill, L.C., N.C. Waran and R.J. Young, 1996. The effect of a foraging device (a modified Edinburgh foodball) on the behaviour of the stabled horse. Appl. Anim. Behav. Sci. 48 (1-2):25-35.

Three-dimensional design of a horse stud like better toll for technical choices of housing and welfare

N. Miraglia and A. Simoni
Molise University, Dept. of Animals, Vegetables and Environmental Sciences, Via De Sanctis, 86100, Italy; miraglia@unimol.it

Abstract

It is well known that captivity and the following upsetting of the horse natural habits and physiological activities have made the horse defenceless towards the numerous dangers and have exposed it to a wide range of risks. As a consequence, the safety of the infrastructures where the athlete spends the majority of its life time represent a fundamental aspect strictly linked to the prevention modalities of many pathologies found within equestrian veterinary practice. In recent year the three-dimensional designs are more and more utilised to have a complete vision of the structures. In the case of horse studs, the three-dimensional design represents the best solution to verify, in virtual imagines, the technical choices not only for housing bur also for improving welfare conditions. In fact it is possible to hypothesise and to virtually test the operative conditions needed both in terms of structures and welfare. The results of this kind of representation is optimal to research the best welfare conditions. This study refers to a virtual three-dimensional design of an equestrian centre destined to show jumping horses. This facility includes the realisation of different structures such as stud and wasting areas, covered grounds, open air fields, open fields, rings, club services, competition arenas with tribunes, jury, services for public, etc.

Keywords: horses, plans, virtual reality, studs

Introduction

Any design of structures destined to horses must keep in mind the horse's basic nature. Fences or stalls adequate for cattle could be hazardous or unsatisfactory for horses. The primary consideration for any horse facility is safety(Warren Evans and Borton, 1990). The horse welfare is the basic aspect in developing a horse farm. Facilities must be designed to protect the horse. Many injuries result from human negligence, poor planning and the facilities in which the horses are kept. Many injuries can be prevented by proper design and a careful construction. Hazards can be eliminated if the horse welfare and safety are the major consideration factors in planning horse facilities. It's well known, as a fact, that exogenous stressors for horses have to be considered boxes dimension, stud temperature, humidity and air flow ventilation (Yaras *et al.*, 2003). Great considerations are actually dedicated also to the geographic orientation and space position of the stud and to their environmental location, without forget to avoid the so called 'novelty factors'(Hada *et al.*, 2003). This means to maintain so long so possible the regularity and intensity of feeding, grooming working, training and resting of the housed horse.

The keys of a successful horse farm are the determination of the goals of the farm and the functionality of the facilities (strictly functional, elaborated studs, etc.); these two elements agree a planning of satisfactory facilities. The programming of the structures for equestrian sports should regard: new interventions, interventions on the existing patrimony(resumption and more rational employment) and the regular and suitable ordinary maintenance of the structures(Donigaglia, 1992). It is not possible to proceed through attempt or adventures. It is necessary to start from a careful exam of the explicit needs of: horse welfare and prevention of injuries, users and the potentialities of the interested area (environment, organisation, social factors, etc.).

Many existent structures were created thanks to the initiative of 'singles': this situation often determined a general lacks in the programming of intervention. In fact some riding centres are characterised by the absence of programming and functional design criteria. The preliminary design must provide conditions and characteristics allowing simple and functional conduction. It is then necessary a previous study of the specific needs and of analogous realisations. In this way it will be possible to provide a design flexible and rational accessible to the customer.

The individuation of some basic characteristics of the intervention will include a time-planned investment program with the definition of a model with the possibility of development in subsequent phases (e.g. planning of a covered riding area to realise in a second step) and the definition of correct functional relations of the designed spaces. The riding centres are generally divided in 3 fundamental sections: sporting activity; services for riders and employees; public. More recently horses and their needs individuate a new section which assumes a specific importance design wise as well as in managerial terms(Angioni *et al.*, 1995).

The equestrian structures are different from all the other sporting centres because there is a join presence of man and horse with presence of differentiated functions linked to areas destined to horses, areas destined to riders and some common areas. Consequently the individuation of necessities concerns the needs of the horses and those of the riders. In the first case the spaces for the horses include studs, enclosed areas, pastures; the spaces with the horses include riding and covered ground etc., storage places(feedstuffs, bedding materials, etc.), shoeing, toilets, first aid, etc. In the second case we refer to dressing rooms, services, first aids, restoration, information, secretariat, etc. The equestrian centres typologies include: structures located in urban areas, often linked to other sporting activities, structures located in rural areas often connected with agricultural activities, centres connected with agritourist and vacation activities, specialised centres destined to equestrian sport activities (show jumping, dressage, three days events).

Traditionally the planning of a riding centre has been always executed from the planners realising designs from which the structures are come true effectively. The traditional design planning includes two projection size designs constituted by plans, prospectus and sections. On the other side the modern design planning include paper designs in two and three dimensions, virtual representation of buildings and figures animation (Simoni, 1998; 2002).

The aim of this paper is to describe the planning of a riding centre destined to show jumping horses with traditional designs and with a 3-D program joined on animations to show the several advantages of the use of virtual reality and animations in planning a riding centre.

Material and methods

The design technique includes the planning of an equestrian facility destined to 40 horses completed by the different buildings and accessories typical of this kind of facility(Angioni *et al.*, 1995; Donigaglia, 1992, Simoni, 1998). The design realisation was realised directly by the CAD and in 3-D Design: Autocad 2000 program in 3 D developed with program 3D max 2,5 (Bacigalupi, 1998; Peterson, 1998; Leach, 2000). The following step was the passage from this file to a program with which it was possible to elaborate virtual animations that agree to the observer to 'move' in the equestrian centre in a virtual reality.

Description of facilities

The equestrian centre hypothesised by this project refers to a facility for 40 horses developed on a surface of 6ha. It includes indoor and outdoors surfaces.

Indoor areas

- Stud: 40 boxes(3.5x3.5m); 4 boxes destined to toilet, washing(solarium) with in-out entrance; saddlery; concentrates storages; directors and instructors office; dressing room for staff (A);
- indoor riding ground: internal surface: 50 x 36 m, realised in lamellar wood (B);
- restaurant and welcome area: octagonal building 140m^2; it agrees to welcome 70 seated persons (C);
- offices and dressing areas for riders: directly connected to the restaurant building (D);
- parking for trucks and trailers: covered lodge 45 x 6 m; closed to the indoor riding ground with special in-out course (E);
- garage: 25x15m, destined to jumps accommodation, agricultural machinery and maintenance shop (F);
- storage: 25x15m, destined to hay and bedding material (G);
- farriery: 25x15m, constituted by opened and closed areas (H);
- horse infirmary: 4 isolated box destined to quarantine (I);
- guardian residence (L).

Outdoor areas

- Outdoor riding ground in grass with tribune: 100x100m destined to show jumping competitions (M);
- outdoor riding ground in sound: 75x70m (N);
- racetrack in sound: 10x180m long 10x120m large, placed around the riding arena in grass and the paddocks (O);
- paddocks: 3 paddocks 35x31; 3 paddocks 52x33m located inside the racetrack (P1-P2);
- ring destined to free jumping: 5x64m long; 5x20 large (Q);
- rings: 2 rings of 15 m diameter each, the 2nd destined to automatic roping (R1-R2);
- vehicle parking: 2000m^2 (S);
- dung hill: 300m^2 (T);
- other external areas: destined to internal roads, horse and pedestrian pathways, areas destined to mobiles prefabricated boxes in case of participants attending show jumping competitions.

Results

Figure 1 shows the general plan of the facility using the traditional design on paper. From this plans it is possible to observe 2 entrances: the main entrance and the lateral entrance destined to tracks,

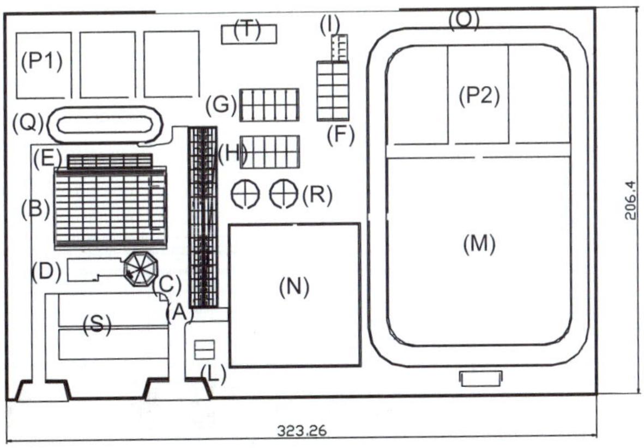

Figure 1. Plan riding centre for 40 horses.

trailer etc. Near the main entrance there is the guardian residence and the porter's lodge. The guardian residence is followed by the stud, on the right, and the restaurant together with offices and services on the left. Close to the stud there is the indoor riding ground with the covered lodge destined to track and trailers. On the back there are the ring destined to free jumping and 3 paddocks. On the right of the stud there are the outdoor riding field, the two rings(including the automatic roping), the farriery area and the hay storage place. Close to the hay storage place there are the garage destined to jumps accommodation agricultural machinery and maintenance shop and the horse infirmary. On the right of the plan there is the race track that include the outdoor riding arena destined to competitions and 3 paddocks. Figures 2 shows the equestrian centre design in 3D. This part is presented also in virtual reality and animations. The following figures show the plans details of the different elements of the facility:

- Figures 3 - 4 - 5: plan, prospectus and section of the stud. These figures represent the traditional designs that agree to evaluate the position of the different places together with the overall dimensions.
- Figures 6 - 7 - 8: 3D presentation of the stud. It agrees to have many views of the stud(outside, inside, differentiation of the places inside the stud(box, horse food preparation, etc.). The animation agrees to have information concerning the horse position inside the stud and, consequently, to verify the correctness of dimensions in relation to animal welfare.

Other examples of the 3D technique design are shown in Figure 9 (restaurant - bar; indoor riding ground, services) and Figure 10 (internal view of the indoor riding ground).

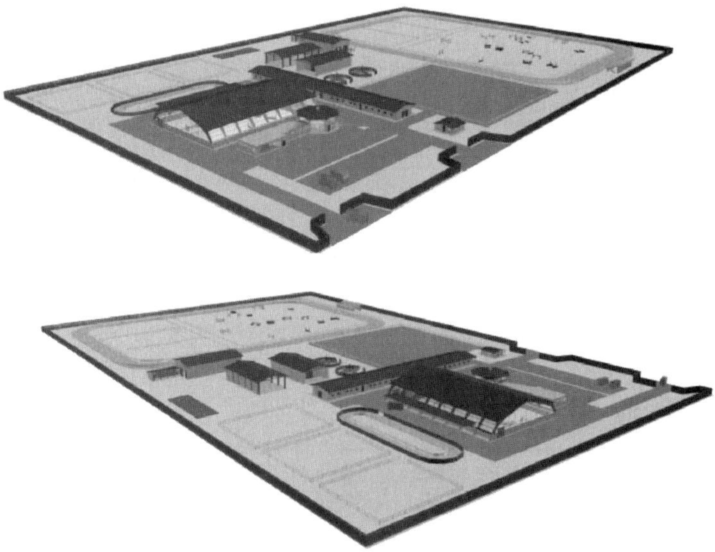

Figure 2. Equestrian centre design three dimensional and virtual reality.

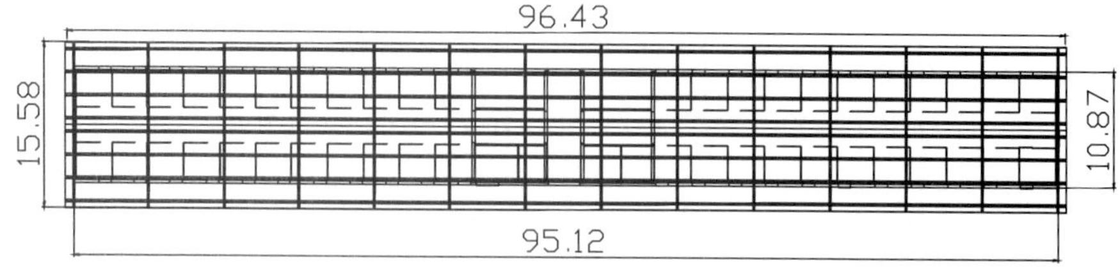

Figure 3. Plan of the studs.

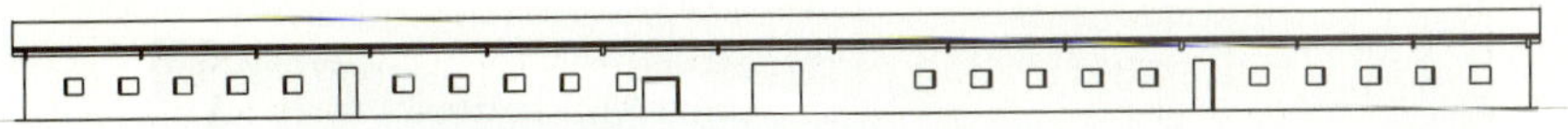

Figure 4. Prospectus of the studs.

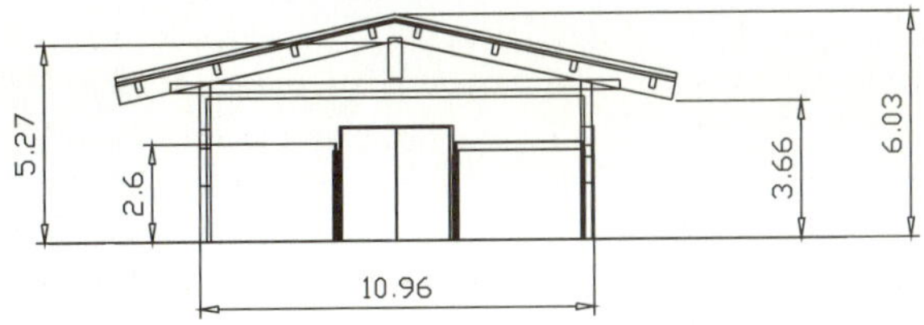

Figure 5. Section of the studs.

Figure 6. Outside view of studs.

Figure 7. Internal view of studs.

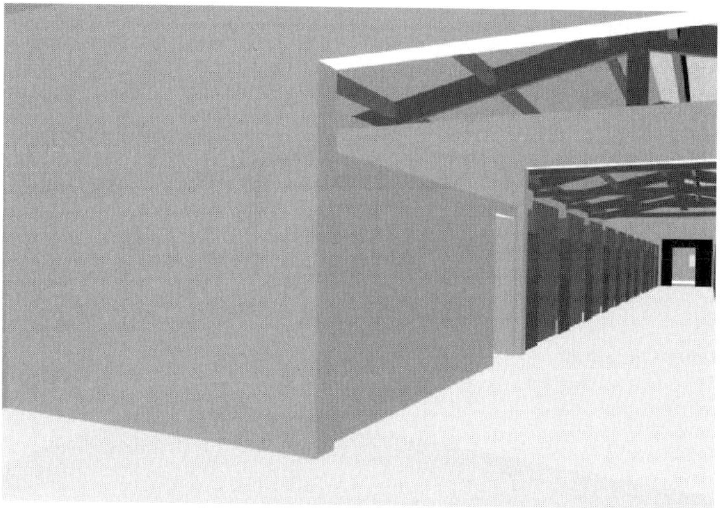

Figure 8. View of horse-food preparation place inside the studs.

Figure 9. Resturant - bar and indoor riding area with grand-stands provided for the public and with relative services.

Figure 10. Internal views of the large covered indoor riding round of about 1800 m² and services.

Discussion and conclusion

The results of this kind of paper are not scientific. They underline, by the observation of the different realisations of the plans, the considerable innovative possibilities coming from this type of program. Obviously the paper description is not sufficient to well appreciate the 3D version and the animations. It is necessary to complete the traditional plan with the computer program. In recent years the three-dimensional designs are more and more utilised to have a complete vision of the structures. In the case of horse facilities, the three-dimensional design represents the best solution to verify, in virtual imagines, the technical choices not only for housing bur also for improving welfare conditions. This kind of technique is useful also to present the project to the customers and it agrees to realise modifications of plans in real time with consequent gain of time and budget.

The use of the 3D planning supported by animations agrees to limit mistakes in the project realisation: this is fundamental when it is necessary to respect and optimise some basic aspects concerning the horse welfare. In fact it is possible to hypothesise and to virtually test the operative conditions needed both in terms of structures and welfare. The individuation of possible planning mistakes will limit the risk of modifications after the realisation of the facility with considerable advantages for horses, customers, users and for designers.

Acknowledgements

Authors wish express their acknowledgement to Daniele Gagliardi DVM, President of the Molise Region Committee of Italian Equestrian Federation (FISE) and FEI Veterinarian for his support in reviewing the technical aspects concerning the facilities destined to Equestrian Sports.

References

Angioni, P., M. Beta, G. Brunetti, E. Carbone, C. Croce, L. Novo, A. Paolucci, M. Reitano, M. Turner, C. Vagnozzi and U. Mezzani, 1995. Equestrian sports facilities. Spaziosport, 1, 25-104.

Bacigalupi, D., 1998. 3D Studio MAX 2.5.

Donigaglia, P., 1992. Studio per una struttura di addestramento per equini da concorso ippico. Thesis, Bologna University.

Hada, T., T. Hanaka, T. Takahashi, A. Hiraga and K. Yagi, 2003. Effects of novelty stress on neuroendocrine activities and roming performances in Thoroughbred horse. J. Endocrinology, 15, 638-648.

Leach, J., 2000. Autocad. McGraw-Hill Ed.

Peterson, M., 1998. 3D Studio MAX 2. Apogeo srl.

Simoni, A., 1998. La progettazione dei ricoveri zootecnici con sistemi di grafica e animazione tridimensionali. Estimo e Territorio, 7/8, 21-24.

Simoni, A., 2002. La progettazione degli allevamenti con grafica tridimensionale, come fattore migliorativo delle scelte tecniche imponibili all'impianto. Atti IV° Convegno 'Nuove acquisizioni in materia di ippologia', Campobasso, 103-111.

Warren Evans, J. and A. Borton, 1990. Fences, buildings and equipment. In: The Horse, J. Warren Evans, A. Borton, H.F. Hintz, L. Dale Van Vleck(editors), Freeman and Company Ed., New York, 753 - 785.

Yaras, N., A. Agar, S. Gumuslu and I. Ordemir, 2003. Effects of immobilization and cold stress on visual evoked potential. Intern. J. Neuroscience, 113, 1055-1067.

Behavioural repertoire: its expression according to environmental conditions

H. Benhajali, M. Hausberger and M.A. Richard-Yris
'Ethology-Evolution-Ecology', Rennes 1 University, UMR CNRS 6552, France

Abstract

The domestic environment is often very different from the natural one. This difference involves differences in the behavioural repertoire of animals that can take different forms. Variations in animals' behavioural repertoire could be qualitative (absence of some behaviours or postures; abnormal behaviours) or quantitative (time budget modifications, abnormal behavioural sequences, modifications of the relative quantities of different acts in a repertoire). These variations can be useful for perceiving the effect of husbandry practices on animals' behaviour and thus their subsequent welfare.

Keywords: behavioural repertoire, time budget, domestic environment, welfare

Introduction

An ethogram is a thorough inventory of all the discrete behaviours every species member can display regardless of gender or age. The usefulness of this tool is indisputable since it provides definite descriptions of the different acts and therefore useful objective criteria for different observers. The notion of behavioural repertoire is more restrictive since it concerns only a part of the ethogram of the species but its interest is that it takes into account the existence of inter-individual differences inside the same species.

The behavioural repertoire of animals changes in response to environmental changes (Inglis and Langton, 2006). Thus changes in animals' environment caused by domestication involved many changes in their behavioural repertoire compared with that of their wild relatives. These behavioural variations can be qualitative (absence or presence of some behaviours) or quantitative (in frequency or temporal repartition of behaviours). The aim of this paper was to review the different kinds of behavioural variations seen in different domestic animals including horses for which we still lack data about the effect that many environmental factors may induce. This review is followed by an example of the use of behavioural repertoire and time budget of horses as tools for assessing their welfare.

Qualitative variations

Restricted behavioural repertoire

Domestic animals still possess the ability to show the full behavioural repertoire of their wild ancestors (Weinert *et al.*, 2001, Mignon-Grasteau *et al.*, 2005). Yet, their behavioural repertoire is often restricted compared with their free-ranging relatives (Dantzer, 1986). This restriction is due to certain captivity conditions which prevent animals from expressing some natural behaviours seen in the wild, either directly (e.g. physical restraint) or indirectly (e.g. absence of particular conspecifics).

Space restriction

To demonstrate natural behaviour, animals need to be in an environment where they can move freely. This freedom is not always provided in the domestic environment where, conversely to their wild relatives, domestic animals are often reared in closed enclosures with limited space.

Space restriction may lead to the absence of some behaviours in domestic animals. Laying hens reared in battery cages can not perform behaviours that require much space, like grooming, dust-bathing and wing-flapping. Rabbits and gerbils housed in small cages can not perform rearing; pregnant mares used in the pregnant mare urine industry (PMU) and tethered in tie stalls for 5-6 months of the year can not display normal locomotion (Jongman *et al.*, 2005). In dairy calves kept in pens, the lack of sufficient space may inhibit play behaviour (Jensen and Kyhn, 2000).

Space restriction may also lead to the absence of some postures. In Le Neindre *et al.*, (1991 In Le Neindre, 1993) () observed lying postures of male Friesian veal calves housed in crates of three sizes (0.55 m x 1.50 m,.65 m x 1.65 m, and 1.10 m x 1.50 m, n = 10 calves per group). All calves were housed on slatted floors and were tethered for 8 wks after their arrival on the farm. Calves reared in small pens (0.55 m x 1.50 m and 0.65 m x 1.65 m) did not lie down with two or all four legs stretched, whereas calves in larger pens (1.10 m x 1.50 m) were able to lie in a normal position (Le Neindre *et al.*, 1991; In: Le Neindre, 1993).

Lack of stimuli or environmental elements

Besides space restriction, the absence of adequate stimuli or some environmental elements in captivity may hamper the expression of some natural behaviour. For example, many species of animals build nests, both for the raising of young as well as for resting and shelter. Absence of adequate substrate in some barren captive environments prevents animals from nest building. For example, pregnant sows kept in farrowing crates, hens and rodents reared in barren cages with no access to nesting material cannot perform nesting. The inability to do so may result in aberrant behaviour (Morgan and Tromborg, 2007). Other examples can be digging behaviour in hamsters and gerbils that can be inhibited by the absence of a suitable digging substrate and perching in poultry which is prevented by the absence of perches.

Intensive animal husbandry practices

In conditions of intensive animal husbandry, some natural behaviours are often neglected. This article is about those behaviours that have no obvious goal and therefore which are thought to be of no interest for animal production. This is the case of appetitive behaviour which is often neglected unlike that of consommatory behaviour. Food restriction and confinement in barren housing prevents captive horses from performing foraging which is predominant activity in the natural state (Goodwin *et al.*, 2002). Another extreme example is intensively managed mating in horses, which prevents them from performing the pre-copulatory behaviour displayed in free mating sequences. In fact in free ranging horses, when a mare is in oestrus, the frequency of interaction of the stallion and a mare increases to as many as hundreds of such encounters and communications per day, with precopulatory 'teasing' sequences as often as every few minutes. In contrast, in intensive domestic breeding, the active role of the mare is limited by management practices. The mare is restrained with a lip or ear twitch, or/and a foreleg lifted to ensure immobility before the breeding stallion is brought to the mare. Handlers of domestic stallions often direct the stallion to approach the mare directly to the hind-quarters, limiting the stallion's access to the head and fore body of the mare and thus limiting the pre-copulatory contact with mares (McDonnell, 2000).

Absence of particular conspecifics

Social environment in domestic conditions is generally very different from that in the natural state. Horses are mostly housed singly or reared in controlled groups with animals of the same sex or age. Absence of some particular conspecifics may affect the behavioural repertoire of domestic animals. Social repertoire can not be performed in the absence of conspecifics. Isolated animals can not express any social behaviour. The absence of sexual partners prevents sexual behaviour and mothers separated from their young do not exibit maternal behaviour.

But besides this direct effect on animals' behavioural repertoire, the absence of some particular conspecifics may hamper the development of normal behavioural patterns since the experience acquired from the mother of other relatives is also important to develop natural behaviour. A good illustration of this can be the study conducted by Bourjade *et al.* (2006) showing that young horses raised without adults showed a restricted behavioural repertoire as compared to data on feral groups. The introduction of adults into the group of young horses induced the expression of new behaviours like lying down, flehmen, mounting, snapping, head bowing, faecal male displays, and allogrooming (Bourjade *et al.*, 2006).

Redirected behaviours

Behavioural studies showed that captive animals need to perform particular behaviour even when their physiological needs are taken care of, for example: foraging behaviour in horses (Goodwin *et al.*, 2002) and primates (Blois-Heulin and Jubin, 2004), rooting behaviour in pigs (Hughes and Duncan, 1988; Horrell *et al.*, 2001), and digging behaviour in mice (Sherwin *et al.*, 2004). These behaviours, commonly called 'behavioural needs' were defined by Hughes and Duncan (1988) as 'internally driven behaviours'.

When restricted (by physical restraint or lack of adequate substrate), some strongly motivated behaviours that would normally be performed towards particular substrates may be directed towards other resources (Wood-Gush and Vestergaard, 1989). For example, tail biting would be a redirected foraging behaviour that corresponds to behaviour prior to feeding. In the wild, pigs will spend over 20% of their time rooting during daylight hours (Stolba and Wood-Gush, 1989). Pigs fed on concentrates and prevented from expressing much appetitive foraging behaviour towards appropriate substrates (e.g. soil) may redirect this behaviour toward pen-mates and tail bite (Wood-Gush and Vestergaard, 1989).

In piglets, massaging the udder and sucking the teats of the sow if missed may be redirected towards other piglets and lifeless objects (Van Putten and Dammers, 1976). Cross-sucking (sucking of ears, tails, prepuce and other body parts) and other types of non-nutritional sucking in calves appear to be caused by a need to suck that has not been satisfied and is thus redirected to inadequate substrates (De Passillé, 2001). Feather pecking is thought to be a form of redirected behaviour, developed from either food pecking, ground pecking or pecking during dust-bathing (Rodenburg and Koene, 2003). Mongolian gerbils develop extensive stereotypic digging, if they are not provided with an artificial nest with a tube-like entrance or are not allowed to dig a proper nest (Wiedenmayer, 1997). In mice, the development of bar biting seems to be closely related to the relevant behaviours such as exploration directed towards the external environment (Nevison *et al.*, 1999).

In stabled horses, bed eating may represent redirection of grazing behaviour without access to a high-fibre forage (Mills *et al.*, 2000). Cooper *et al.* (2005) showed that increasing meal frequency decreased bed directed behaviour in stabled horses. Goodwin et *al.* (2002) found that providing horses with more than one source of forage decreased the duration and frequency of bed foraging in stabled horses (Figure 1; Goodwin *et al.*, 2002).

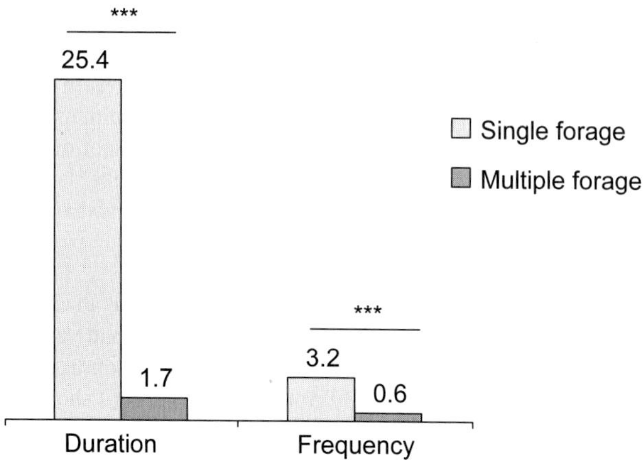

Figure 1. Mean duration and frequency of bed foraging in stabled horses provided with single or multiple forages (Goodwin et al.*, 2002).*

In some captivity conditions, absence of a sexual partner may lead animals to redirect their sexual behaviour toward individuals of the same sex. Male guppies *Poecilia reticulata,* direct sexual behaviour towards other males when housed in single-sex tanks (Field and Waite, 2004).

Quantitative rather than qualitative variations

Even if as we saw above, intensive housing conditions may lead to an impoverishment of animals' behavioural repertoire, bigger variations are often visible when we look at behaviour frequencies and their temporal repartition or the organisation of the sequences of acts expressed by animals according to their environment.

Time budget modifications

Boyd and Keiper (2005) defined Time budget as the amount of time invested by animals in each of their activities. Time budget varies with the environment and since a domestic environment gives a different home range to animals, their time budget may be quite different. Housing and management practices influence considerably the domestic animals' time budget.

Effect of feeding practices

Many animals spend a large portion of their daily activity budget in the search for and consumption of food in their natural habitats (Herbers, 1981). Domestic environment is frequently associated with a restricted access to forage, low fibre diet with concentrates and pellets that can be rapidly consumed by animals. Domestic animals spend thus less time feeding than their free-ranging counterparts.

Diet affects patterns of sleep and wakefulness in horses. Dallaire and Ruckebusch (1974) found that the total recumbency time of housed ponies was increased by about 20% when oats were substituted for hay in their diet. Diet may influence time stand resting, maybe by modifying the time spent feeding, the latter was found to be inversely related to time spent grazing (Duncan, 1980). Food restricted pigs spent less time lying and changed posture to lying less often than pigs fed ad libitum (Day *et al.*, 1995).

Feeding practices may also influence the time spent in locomotion. Restricting the feeding of captive giraffe in zoos resulted in increased locomotor activity compared to free-ranging giraffe (Veasey *et al.*, 1996). Foraging enrichment of stabled horses decreases the time they spent moving around the stable (Goodwin *et al.*, 2002).

Feeding practices may also affect the social behaviour of domestic animals. Several studies showed that foraging enrichment reduced aggressions in group housed animals in primates (Chamove *et al.*, 1982; Chamove and Anderson, 1979; Davys, 1995; Anderson and Chamove, 1984; Chamove *et al.*, 1984; Boccia, 1989a; Boccia, 1989b); pigs (Sneddon and Beattie, 1995; Wood-Gush and Beilharz, 1983; Appleby and Wood-Gush, 1988; Blackshaw *et al.*, 1997; Beattie *et al.*, 2000), rats (Johnson *et al.*, 2004) and laying hens (Huber-eicher and Wechsler, 1998). Feeding enrichment for primates increased positive social interactions in some trials (Boccia *et al.*, 1995; Ludes-Fraulob and Anderson, 1999; Ludes and Anderson, 1996) but decreased them in other trials (Byrne and Suomi, 1991; Brent, 1992; Blois-Heulin and Jubin, 2004).

The decrease of time spent feeding has been associated with stereotypies and abnormal behaviours in horses (Willard *et al.*, 1977; Broom and Kennedy, 1993; McGreevy *et al.*, 1995a), pigs (Appleby and Lawrence, 1987, Young *et al.*, 1994), primates (Marriner and Drickamer, 1994; Bayne *et al.*, 1991), captive giraffe and okapi (Bashaw *et al.*, 2001). Conversely, providing animals with foraging opportunities increases the time spent foraging and reduces abnormal behaviours reporting the animal's time budget closer to the natural state (Young *et al.*, 1994; Baker, 1997; Goodwin *et al.*, 2002; Thorne *et al.*, 2005; Goodwin *et al.*, 2007; Blois-Heulin and Jubin, 2004).

Effect of social environment

Most of the domestic animals are social species, as sociality is one of the characteristics of a species that facilitates the process of domestication (Price, 1984). However, domestic animals are often singly housed and have little opportunity for contact with their conspecifics.

Social isolation is a disturbing experience for social animals. Separation from conspecifics increases respiration rate, serum cortisol, vocalisation, and locomotion in sheep (Carbajal and Orihuela, 2001), cattle (Boissy and Le Neindre, 1997), and horses (Mal *et al.*, 1991) and therefore used to induce stress in many research investigations (Hausberger *et al.*, 2004).

The lack of social contact has been shown to influence horse's time budget. Comparing horses housed in groups to those housed alone, Kiley-Worthington (1984) found solitary individuals spent less time eating and more time standing as well as sleeping than group-living horses. Heleski *et al.* (2002) compared the behaviour of horses weaned individually in solid partition box stalls versus horses weaned in small groups and housed in paddocks. Paddock housed weanling horses spent significantly more time moving and less time lying and engaged in aberrant behaviour than stalled weanlings (Table 1; Heleski *et al.*, 2002). Christensen *et al.* (2002) compared the behaviour of previously group stabled horses and previously singly stabled released in into two separate 2 ha enclosures according to the treatment. More aggressive behaviour was recorded in the group of previously singly stabled stallions whereas group stabled stallions tended to make more use of subtle agonistic interactions (displacements, submissive behaviour).

The lack of social contact has been associated with the occurrence of abnormal behaviours (Cooper *et al.*, 2000; Mills and Davenport, 2002; McAfee *et al.*, 2002). Houpt and McDonnell (1993) reported that providing singly housed stallions with a stall companion reduce the incidence of self directed aggression.

Table 1. Time budget of stall/single versus paddock/group-housed horses (Heleski et al., *2002).*

Behaviour	Paddock/group-housed (Mean ± Std)	Stall/single-housed (Mean ± Std)
Lying down	3.0 ± 0.8%	21.2 ± 3.0%***
Standing	36.1 ± 4.0	24.0 ± 1.5**
Moving around/locomoting	8.3 ± 1.2	4.1 ± 2.3*
Investigating	3.1 ± 2.9	13.0 ± 3.2**
Aberrant	0.4 ± 0.2	4.7 ± 2.6**
Eating hay/grain	23.9 ± 1.9	29.1 ± 2.1^{N}
Other	5.0 ± 0.3	5.0 ± 0.8^{N}
Grazing	18.5 ± 3.4	Not available
Social interactions	3.7 ± 2.5	Not available

Overall time budget differs between the two treatments ($P < 0.0001$). Listed percentage is the percentage of total instantaneous observations.
^{N} = $P > 0.1$ (not significant)
* = $P < 0.1$ (trend toward significance)
** = $P < 0.005$ (significant)
*** = $P < 0.001$ (highly significant)

In other captive situations where animals are maintained in social groupings with high social density, group size or a group composition that would not generally occur in nature (e.g. Stoinski *et al.*, 2001), social contact may be a source of stress and influence the time budget of the animals by increasing aggressions and stereotypies. Increased social density exacerbated aggressions in primates (Elton, 1979; Ha *et al.*, 1999; Plowman *et al.*, 2005) and feather pecking in laying hens (Nicol, 1999). In group housed pigs, Meunier-Salaûn *et al.* (1987) found that severe area restriction decreased social activity. Aggressive behaviour was less frequent in low and high densities than in medium one. Similar results were found in hens by Polley *et al.* (1974).

In domestic horses, effects of size, density and group composition have not been carefully investigated to date. Domestic group sizes exceed generally the most commonly recorded natural group size of four or five individuals (Waring, 1983) and that may theoretically affect the level of interactions within groups. In group-housed dairy calves, group size was found to influence newly weaned dairy calves' time budget and their social interactions. Calves kept in larger groups were more active, spent more time lying and eating close to another calf and performed fewer displacements from the feed-rack (Færevik *et al.*, 2006).

Effect of housing: paddock size, box size, beddings, daily exercise and time spent outside

The box/stall size seems to influence resting patterns in domestic animals. Horses housed in a large box spend more time recumbent than those housed in a small one (Raabymagle and Ladewig, 2006). Pigs in smaller pens spend more time sitting or standing motionless, compared to pigs housed in larger enclosures (Pearce and Patterson, 1993). Jørgensen and Bøe (2007) investigated the effect of paddock size on the behaviour of individually kept horses. A larger paddock size increased time spent eating grass and decreased passive standing. In other studies of group-housed animals, housing the animals in smaller pens increased rates of agonistic behaviour in pigs (Lammers and Schouten, 1985, Wiegand *et al.*, 1994), calves (Napolitano *et al.*, 2004) and turkeys (Buchwalder and Huber-Eicher, 2004).

Bedding influences the time spent lying in cows (Manninen *et al.*, 2002), pigs (Tuyttens *et al.*, 2004) and horses (Pederson *et al.*, 2004). Horses kept on straw bedding spent three times longer lying in lateral recumbency than horses kept on shavings (Figure 2; Pederson *et al.*, 2004). It has been reported also that the use of non-straw bedding may increase the risk of abnormal behaviour and redirected behaviours (De Jong *et al.*, 1998; McGreevy *et al.*, 1995a).

Confinement of horses for more than a given number of hours in the day increases locomotion after release (so-called rebound effect). Houpt *et al.* (2001) showed that when released after confinement for two weeks, non exercised mares showed compensatory increases in locomotion in comparison to mares exercised daily whereas the behaviours in the stalls of exercised and confined mares were similar. Similarly, Jørgensen and Bøe (2007) found that daily exercise decreased not locomotion but also activity in general and reduced the time spent alert standing. It may also influence also the time spent grazing. Chaya *et al.* (2006) compared whether there were any differences in behaviour in stabled horses that were allowed either 2- or 12 hours of turnout time each week. Horses turned out for 2 hours a week trotted, cantered and bucked more frequently and grazed less frequently than horses turned out for 12 hours a week.

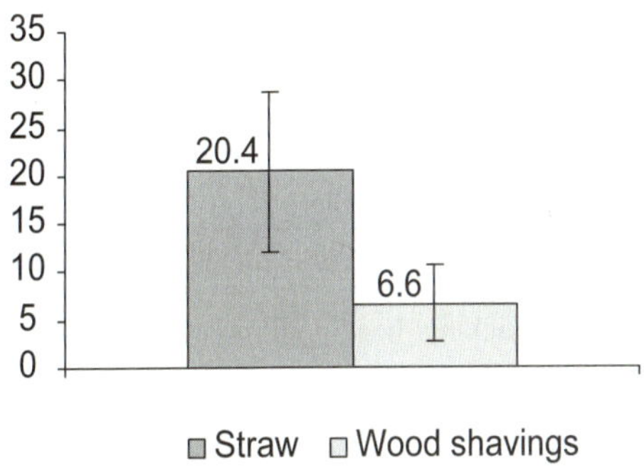

Figure 2. Average longest period of non-interrupted lateral recumbency in minutes (Pederson et al.*, 2004).*

Modifications of the relative quantities of different acts in the repertoire

Another type of quantitative behavioural changes due to the domestic conditions can be the modification of the relative quantities of different acts in the repertoire. This is, for example, the case of ringed gilts. Pigs ringed with bullrings through the nasal septum or clip rings through the upper rim of the snout cannot root (Studnitz *et al.*, 2003a) and therefore try to find other ways to root. The study performed by Studnitz *et al.* (2003b) showed that there was no significant difference between how often the ringed and the unringed gilts were observed foraging or exploring including eating, grazing, chewing, rooting, sniffing, and manipulating. Ringed gilts substituted their rooting behaviour with other exploratory or foraging behavioural elements depending on the possibilities (Studnitz *et al.*, 2003b).

Abnormal behavioural sequences

Some captivity conditions may involve altered behavioural sequences. This is the case of getting up and lying down movements in particular in cattle. These movements are performed according to a native pattern, which can be divided into sub-sequences and described in time consumption. The main environmental restrictions on natural behaviour when performing these movements seem to be: (1) too little space in front of the animal (tied systems and cubicles); (2) slippery floors (cubicles) (Lidfors, 1989).

Krohn and Rasmussen, (1992) found that the time needed for the cows to lie down was longer and the number of interruptions was higher among tied than loose-housed cows. Young cattle on slippery slatted floors show alterations in behaviour which inhibit, delay or prolong lying (Andrae and Smidt, 1982).

The performance of getting up and lying down has been used to evaluate different environments for cattle and as a functional test for the evaluation of limb health of dairy cows kept in dairy environments (Lidfors, 1989).

Another example of abnormal behavioural sequences that can appear in domestic environment is stereotypies defined as repeated, relatively invariable sequences of movements which have no obvious purpose (Figure 3; Cronin *et al.*, 1985). The repetition may be regular or not and the sequence of movements may be very short (e.g. head shaking in hens) or long and complex (e.g. route-tracing by bears, stall walking by horses).

Stereotypies are typical of animals in some conditions of captivity (reviewed by Mason, 1991). The link between stereotypies and animal welfare is a highly debated issue. For some authors stereotypies are a sign of a sub-optimal environment (e.g. Hediger 1950; Stolba *et al.* 1983), in which the animal's welfare is poor (e.g. Wiepkema 1983; Broom 1983) but for others, increased stereotypy does not necessarily reflect a reduction in well-being (e.g. Mason, 1991).

Examples are bar-biting and tongue-rolling in tethered cows (Krohn, 1994), stereotyped movements in dairy cattle (Friend, 1991) and abnormal digging in Mongolian gerbils (Sørensen *et al.*, 2005). In domestic horses, a variety of stereotypies has been observed such as weaving, box walking, wind-sucking, crib biting, pawing, door-kicking, head nodding and head shaking (Mills, 2005).

Two factors in particular have been associated with the development of stereotypic behaviour in stabled horses: social isolation and the feeding of concentrates without access to high-fibre forage. Cooper *et al.* (2000) found that providing additional opportunities to view or socially interact with other horses reduced the incidence of weaving and head nodding in stabled horses. This effect can be replicated with mirrors (Mills and Davenport, 2002; McAfee *et al.*, 2002). The feeding of high-energy, low-fibre concentrated feeds without access to high-fibre forage is associated with a higher incidence of stereotypic activities (McGreevy *et al.*, 1995a; Nicol, 1999; Gillham *et al.*, 1994; Johnson *et al.*, 1998). Oral stereotypies are more common in horses without access to forage (Willard *et al.*, 1977; McGreevy *et al.*, 1995a; Johnson *et al.*, 1998). Foraging enrichment for stabled horses significantly reduces the incidence of these activities (Goodwin *et al.*, 2002; Thorne *et al.*, 2005).

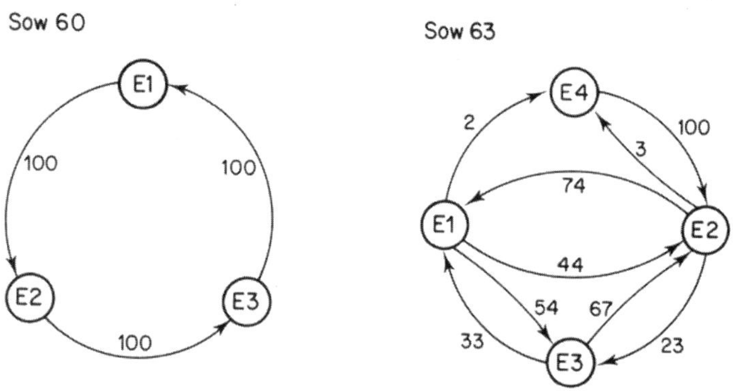

Figure 3. Quantitative description of stereotyped sequences in two pregnant tethered sows. Each element corresponds to a specific motor. Numbers along arrows indicate the percentage of occasions that individual elements succeeded other elements (Cronin et al.*, 1985).*

McGreevy *et al.* studied the prevalence of wood-chewing, weaving, crib-biting/wind-sucking and box-walking in horses competing in different disciplines (dressage, eventing and endurance) (Figure 4; McGreevy *et al.*, 1995b). They found significant linear trends for the eventing group ($p < 0.001$) and the dressage group ($p < 0.05$). It was concluded that the time a horse spends out of the stable is related to the discipline for which it is being trained and in dressage and eventing horses the time spent in a stable is correlated with an increased risk of abnormal behaviour.

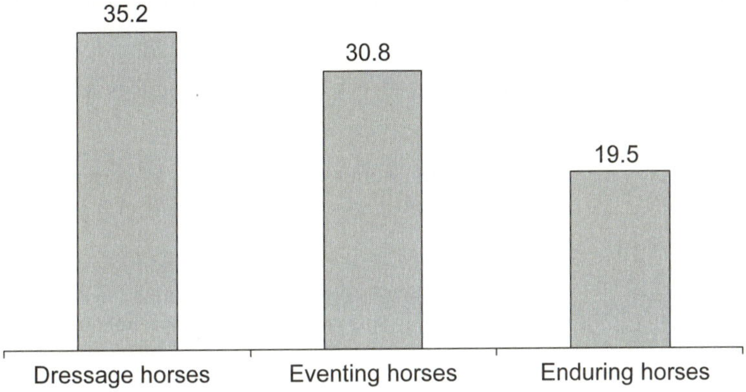

Figure 4. The reported percentage prevalences of stereotypic behaviour for the dressage, eventing and endurance horses (McGreevy et al.*, 1995b).*

Modifications of the daily or seasonally distribution of activities

Animals may change their behavioural rhythm according to their environment features (e.g. food distribution times) or to avoid some particular conspecifics (e.g. subordinate domestic mouse may change their behavioural rhythm to avoid contact with dominants (Crowcroft and Rowe, 1963 In Regal and Connolly, 1979).

These behavioural rhythm modifications can also be imposed by certain intensive breeding practices. Some farm species have their reproductive period during the season where the days are long (long-day breeders). This annual reproductive rhythm is controlled by the length of the daylight. Treatment of females with an artificial light during the seasonal anoestrus, which is a commonly used practice in intensive animal husbandry induce sexual behaviour in horses (Guillaume, 1996), sheep and goats (Thimonier, 1981)

Natural environment and welfare

The natural environment is generally taken as a benchmark to evaluate different domestic environments. The naturalness of the behavioural repertoire has been used as a welfare indicator (e.g. Webster *et al.*, 1986). But do animals need to perform all behaviours seen in the wild or is their welfare always good within the natural environment?

In the natural environment, animals experience disease and injury and may be subject to predation, all of which reduce welfare. The performance of aspects of the wild or natural behavioural repertoire, such as escape and frustration-related behaviours, clearly does not reflect an animal having a good state of psychological welfare (Hughes and Duncan, 1988, Spinka, 2006). Thus, Absence of some natural behaviour in captivity is not always a negative sign of welfare. Natural behaviours can be classified into those whose absence in the domestic environment reduce animal welfare and those whose absence has no importance/or advantage (predator-flight or competition for resources) regarding the animal welfare.

Behavioural repertoire and time budget as useful tools for assessing domestic animals' welfare: a case study in domestic horses

We recently conducted an experiment aiming at investigating the effect of foraging opportunities on the behavioural repertoire, time budge and social behaviour of Arab breeding mares and their subsequent welfare. The experiment was conducted at the national breeding facility of Sidi Thabet, located at 20 km from Tunis, Tunisia. In the Sidi Thabet Stud, mares were housed in individual boxes, where pellets and hay were distributed in the early morning and the evening. They were allowed outside between 9 a.m. and 3 p.m. in a bare paddock where free access to water was provided. Once a day, cut grass was distributed in the paddock but in small quantities, which was generally all consumed by mares in the following hour.

One hundred pure-bred Arab breeding mares, from 4 to 22 years old (all barren) were used in the experiment. None had been bred for the ongoing season at the beginning of the experiment.

The treatment consisted in providing hay to the experimental group. The 100 mares were randomly divided into 2 groups of 50 mares: the experimental and the control groups and separated during the day in 2 bare paddocks at the same density. Hay was provided in the experimental paddock in nylon haynets. These were filled with hay and hung up in the experimental paddock every morning before the arrival of mares and taken down every evening once the mares had left the paddock. There were as many haynets as mares. A distance of 3 m was respected between the haynets. Twenty-minute animal focal samplings and scan samplings were used to determine the time budget of the mares during the period from 9 a.m. to 3 p.m. and study their social behaviour.

A total of 300 focal sampling (6000 min), 3300 individual scan sampling (6000 min) and 62 group observations (1240 minutes) corresponding to 100 mares were recorded. Results showed that the hay treatment involved qualitative and quantitative behavioural variations between the two groups.

Behavioural repertoire

From recorded behavioural observations, we drew-up the behavioural repertoire of each group of mares which listed all the behaviours performed by mares throughout all the observations.

The experimental group showed a behavioural repertoire which was broader than that of the control one. Some behaviours observed in the experimental group were not observed in the control one. These behaviours were allogrooming and lying down in both the sternal and the lateral position.

Both lying down and allogrooming are important to the horse's well being. A certain level of recumbency is critical to their well being (Boyd, 1998) because paradoxical sleep usually occurs in the recumbent position and is thought to be important in memory sorting and consolidation (Dallaire, 1986). Allogrooming is indicative of companionship between individuals (Kimura, 1998) which is now being recognised in horse-behaviour research as important to the well being of domestic horses (Feh, 2005).

Time budget

Time budgets of experimental and control mares showed significant differences between the two groups. Experimental mares' time budgets were closer to the natural state with significantly more time spent feeding, less time spent in locomotion, stand resting and alert standing (Table 2).

Table 2. Average time budget of experimental and control groups (Mean ± S.E.)

	Control group Mean ± S.E. (%)	Experimental group Mean ± S.E. (%)	*p*-value
Feeding	29.75 ± 2.45	65.12 ± 2.40	$p < 0.01$
Stand resting	27.52 ± 2.62	11.76 ± 2.57	$p < 0.01$
Locomotion	23.56 ± 1.34	11.70 ± 1.31	$p < 0.01$
Alert standing	14.71 ± 1.23	5.23 ± 1.2	$p < 0.01$
Self grooming	0.93 ± 0.25	0.95 ± 0.25	$p > 0.05$
Rolling	0.18 ± 0.08	0.06 ± 0.08	$p > 0.05$
Drinking	1.42 ± 0.37	2.02 ± 0.37	$p > 0.05$
Social interaction	1.17 ± 0.32	1.66 ± 0.31	$p > 0.05$
Defecation	0.25 ± 0.15	0.47 ± 0.15	$p > 0.05$
Urination	0.18 ± 0.08	0.06 ± 0.08	$p > 0.05$
Others*	0.62 ± 0.46	1.90 ± 0.45	$p > 0.05$

*Other behaviour

The decrease of time spent in locomotion by experimental mares may be explained by a lowered level of stress, as locomotion increases in stressful situations (Carbajal and Orihuela, 2001; Houpt, 2005).

Social behaviour

The frequency of social interactions observed during individual focal samplings was significantly higher in the control group (Table 3) but this is rather due to a higher frequency of agonistic interactions in the control group ($p < 0.01$) (Table 3). However, the frequency of positive social behaviour was significantly higher in the experimental group ($p < 0.01$) (Table 3).

Results obtained showed also that there was significantly more preferential associations in the experimental group than in the control one (26 vs. 12, Chi square = 5.5384, $p < 0.05$).

According to the findings of this study, it seems that the restriction of some behaviours (caused by the lack of stimuli or physical restraint) may lead to the restriction of other categories of behaviour even if the stimulus for these latter is available. To our knowledge, this study is the first to show that the lack of foraging opportunities influenced not only time budget and behavioural repertoire of horses but may also influence their social behaviour.

Table 3. Frequencies per mare per hour of social interactions, agonistic interactions, positive social interactions and allogrooming observed during focal samplings.

	Experimental group Mean ± S.E. (%)	Control group Mean ± S.E. (%)	*p*-value
Social interactions	9.3 ± 7.17	15.67 ± 18.25	$p < 0.05$
Agonistic interactions	8.51 ± 6.86	15.42 ± 18.26	$p < 0.05$
Positive social interactions	0.79 ± 1.86	0.24 ± 0.89	$p < 0.05$

Conclusion

Animal behaviour changes in response to environmental changes and can provide us with information about how animals react in response to captivity conditions. While keeping in mind the capacities of adaptation of animals, this information can be used in assessing domestic animals' welfare. However a better knowledge of the effect of the domestic environment on animals' behaviour is required to be able to evaluate different housing and husbandry practices.

References

Anderson, J.R. and A.S. Chamove, 1984. Allowing captive primates to forage. In: Standards in Laboratory Animal Management, Part 2, The Universities Federation for Animal Welfare, UFAW, Potters Bar, UK, 253-256.

Andrae, U. and D. Smidt, 1982. Behavioural alterations in young cattle on slatted floors. In: W. Bessei (Ed.) Disturbed behaviour in Farm Animals. EEC- seminar. Hohenheimer Arbeiten, Heft 121, 51-60.

Appleby, M.C. and A.B. Lawrence, 1987. Food restriction as a cause of stereotypic behaviour in tethered gilts. Anim. Prod. 45,103-110.

Appleby, M.C. and D.G.M. Wood-Gush, 1988. Effect of earth as an additional stimulus on the behaviour of confined piglets. Behav. Processes.17, 83-91.

Baker, K.C., 1997. Straw and forage material ameliorate abnormal behaviors in adult chimpanzees, Zoo. Biol. 16, 225-236.

Bashaw, M.J., L.R. Tarou, T.S. Maki and T.L. Maple, 2001. A survey assessment of variables related to stereotypy in captive giraffe and okapi, Appl. Anim. Behav. Sci. 73, 235-247.

Bayne K, H. Mainzer, S. Dexter, G. Campbell, F. Yamada and S. Suomi, 1991. The reduction of abnormal behaviours in individually housed rhesus monkeys (*Macaca mulatta*) with a foraging/grooming board. Am. J. Epidemiol. 23, 23-35.

Beattie, V.E., N.E. O'Connell and B.W. Moss, 2000. Influence of environmental enrichment on the behaviour, performance and meat quality of domestic pigs. Livest. Prod. Sci. 65, 71-79.

Blackshaw, J.K., F.J. Thomas, and J. Lee, 1997. The effect of a fixed or free toy on the growth rate and aggressive behaviour of weaned pigs and the influence of hierarchy on initial investigation of the toys, Appl. Anim. Behav. Sci. 53, 203-212.

Blois-Heulin, C. and R. Jubin, 2004. Influence of the presence of seeds and litter on the behaviour of captive red-capped mangabeys Cercocebus torquatus torquatus. Appl. Anim. Behav. Sci. 85, 340-362

Boccia, M.L., 1989a. Preliminary report on the use of a natural foraging task to reduce aggression and stereotypies in socially housed pigtail macaques, Lab. Primate. Newsl. 28, 3-4.

Boccia, M.L., 1989b. Long-term effects of a natural foraging task on aggression and stereotypies in socially housed pigtail macaques, Lab. Primate. Newsl. 28, 18-19.

Boccia, M.L., M.L. Laudenslager and M.L. Reite, 1995. Individual differences in macaques' responses to stressors based on social and physiological factors: implications for primate welfare and research outcomes. Lab. Anim. 29, 250-257.

Boissy A. and P. Le Neindre, 1997. Behavioural, cardiac, and cortisol responses to brief peer separation and reunion in cattle. Physiol. Behav. 61, 693-699.

Bourjade, M., M. Moulinot, S. Henry, M.A. Richard-Yris and M. Hausberger, 2006. Effet de la présence d'adultes sur le développement du comportement de jeunes chevaux domestiques élevés en groupe. Actes de la 32eme journée de la recherche équine, Paris, 1er mars 2006.

Boyd, L., 1998. The 24-h time budget of a takh harem stallion (*Equus ferus przewalskii*) pre- and post-reintroduction, Appl. Anim. Behav. Sci. 60, 291-299.

Boyd, L. and R. Keiper, 2005. Behavioural Ecology of Feral Horses. In: The domestic horse, the evolution, development and management of its behaviour D. Mills and S. McDonnell (Eds.). Cambridge University Press. 55-81.

Brent, L., 1992. Woodchip bedding as enrichment for captive chimpanzees in an outdoor enclosure. Anim. Welfare. 1, 161-170.

Broom, D.M., 1983. Stereotypies as animal welfare indicators. In: D. Schmidt (Ed.) Indicators Relevant to Farm Animal Welfare. The Hague: Martinus Nijhoff. 81-87.

Broom, D.M. and M.J. Kennedy, 1993. Stereotypies in horses: their relevance to welfare and causation. Equin. Vet. Ed. 5, 151-155.

Buchwalder, T. and B. Huber-Eicher, 2004. Effect of increased floor space on aggressive behaviour in male turkeys (*Meleagris gallopovo*), Appl. Anim. Behav. Sci. 89, 207-214.

Byrne, G.D. and S.J. Suomi, 1991. Effects of woodchips and buried food on behavior patterns and psychological well-being of captive rhesus monkeys. Am. J. Primatol. 23, 141-151.

Carbajal, S. and Orihuela, A., 2001. Minimal number of conspecifics needed to minimize the stress response of isolated mature ewes, J. Appl. Anim. Welfare Sci. 4, 249-255.

Chamove, A.S. and J.R. Anderson, 1979. Woodchip litter in a macaque group. J. Inst. Anim. Tech. 30, 69-74.

Chamove, A.S., J.R. Anderson, S.C. Morgan-Jones and S.P. Jones, 1982. Deep woodchip litter: hygiene, feeding, and behavioural enhancement in eight primate species, Int. J. Stud. Anim. Prob. 3, 308-318.

Chamove, A.S., Anderson, J.R., and Nash, V.J., 1984. Social and environmental influences on self-aggression in monkeys, Primates. 25, 319-325.

Chaya, L., E. Cowan, and B. McGuire, 2006. A note on the relationship between time spent in turnout and behaviour during turnout in horses (*Equus caballus*). Appl. Anim. Behav. Sci. 98, 155-160.

Christensen, J.W., J. Ladewig, E. Søndergaard and J. Malmkvist, 2002. Effects of individual versus group stabling on social behaviour in domestic stallions. Appl. Anim. Behav. Sci. 75, 233-248.

Cooper, JJ, L. McDonald and DS. Mills, 2000. The effect of increasing visual horizons on stereotypic weaving: implications for the social housing of stabled horses. Appl. Anim. Behav. Sci. 69, 67-83.

Cooper, J.J., N. Mcall, S. Johnson and H.P.B. Davidson, 2005. The short-term effects of increasing meal frequency on stereotypic behaviour of stabled horses. Appl. Anim. Behav. Sci. 90, 351-364.

Cronin, G.M., P.R. Wiepkema and J.M. Ree, 1985. Endogenous opioids are involved in abnormal stereotyped behaviours of tethered sows. Neuropeptides 6, 527-530.

Dallaire, A., 1986. Rest behavior. Vet. Clin. North Am. — Equine Practice 2, 591-607.

Dallaire, A. and Y. Ruckebusch, 1974. Sleep and wakefulness in the housed pony under different dietary conditions, Can. J. Comp. Med. 38, 65-71.

Dantzer R., 1986. Behavioral, physiological, and functional aspects of stereotyped behavior: a review and reinterpretation, J. Anim. Sci. 62, 1776-1786.

Day, J.E.L., I. Kyriazakis and A.B. Lawrence, 1995. The effect of food deprivation on the expression of foraging and exploratory behaviour in the growing pig. App. Anim. Behav. Sci. 42, 193-206.

Davys, J.S., 1995. The development of a forage diet for a breeding colony of rhesus monkeys, Anim. Techn. 46, 63-66.

De Jong, I.C., E.D. Ekkel, J.A. van de Burgwal, E. Lambooij, S.M. Korte, Marko A.W. Ruis, J.M. Koolhaas and H.J. Blokhuis, 1998. Effects of strawbedding on physiological responses to stressors and behavior in growing pigs. Physiol. Behav. 64, 303-310.

De Passillé, A.M.B., 2001. Sucking motivation and related problems in calves, Appl. Anim. Behav. Sci. 72, 175-187.

Duncan, P., 1980. Time-budgets of Camargue horses. II. Time-budgets of adult horses and weaned sub-adults. Behav. 72, 26-49.

Elton, R.H., 1979. Baboon behavior under crowded conditions. In: J. Erwin, T.L. Maple and G. Mitchell (Eds.) Captivity and Behavior: Primates in breeding colonies, laboratories and zoos. Van Nostrand Reinhold Company, New York, 139-171.

Færevik G., I.L. Andersen, M.B. Jensen and K.E. Bøe, 2006. Behaviour and social interactions of dairy calves kept at different group sizes. In: ISAE: 40th International Congress of the ISAE, 08.-12.08.2006, Bristol; Proceedings of the 40th International Congress of the ISAE, p. 99.

Feh, C., 2005. Relationships and communication in socially natural horse herds. In: D. Mills and S. McDonnell (Eds.) The domestic horse, the evolution, development and management of its behaviour. Cambridge University Press. 83-93

Field K.L. and T.A. Waite, 2004. Absence of female conspecifics induces homosexual behaviour in male guppies. Anim. Behav. 68, 1381-1389.

Friend, T.H., 1991. Behavioural aspects of stress. J. Dairy Sci. 74, 292-303.

Gillham, S.B., N.H. Dodman, L. Shuster, R. Kream and W. Rand, 1994. The effect of diet on cribbing behaviour and plasma B-endorphin in horses. Appl. Anim. Behav. Sci. 41, 147-153.
Goodwin, D., H.P.B. Davidson and P. Harris, 2002. Foraging enrichment for stabled horses: effects on behaviour and selection, Eq. Vet. J. 34, 686-691.
Goodwin, D., H.P.B. Davidson and P. Harris, 2007. A note on behaviour of stabled horses with foraging devices in mangers and buckets. App. Anim. Behav. Sci. In press.
Guillaume D., 1996. Action de la photopériode sur la reproduction des équidés. INRA Prod. Anim., 9, 61-69.
Ha, J.C., R.L. Robinette and G.P. Sackett, 1999. Social housing and pregnancy outcome in captive pigtailed macaques, Am. J. Primatol. 47, 53-163.
Hausberger, M., C. Bruderer, N. Le Scolan and J-S. Pierre, 2004. Interplay Between Environmental and Genetic Factors in Temperament/Personality Traits in Horses (*Equus caballus*). J. Comp. Psychol. 118, 434-446.
Hediger, H. 1950. Wild Animals in Captivity. London: Butterworths Scientific Publications.
Heleski, C.R., A.C. Shelle, B.D. Nielsen, and A.J. Zanella, 2002. Influence of housing on weanling horse behavior and subsequent welfare. Appl. Anim. Behav. Sci. 78, 291-302.
Herbers, J.M., 1981.Time resources and laziness in animals, Oecologia. 49, 252-262.
Horrell, I., P. A'Ness, S.A. Edwards and J. Eddison, 2001. The use of nose-ringing in pigs: consequences for rooting, other functional activities, and welfare. Anim. Welfare. 10, 3-22.
Houpt, K.A., 2005. Maintenance behaviours. In: D. Mills and S. McDonnell (Eds.) The domestic Horse; the evolution, development and management of its behaviour. Cambridge University Press, 94-108.
Houpt KA and SM. McDonnell, 1993. Equine stereotypies. Compend. Cont. Educ. Pract. Vet. 15, 1265-1272.
Houpt, K.A., T.R. Houpt, J.L. Johnson, H.N. Erb and S.C. Yeon, 2001. The effect of exercise deprivation on the behaviour and physiology of straight stall confined pregnant mares. Anim Welfare. 10, 257-267.
Huber-eicher, B. and B. Wechsler, 1998. The effect of quality and availability of foraging materials on feather pecking in laying hen chicks. Anim. Behav. 55, 861-73.
Hughes, B. and I.J.H. Duncan, 1988. The notion of ethological need, models of motivation and animal welfare. Anim. Behav. 36, 1696-1707.
Inglis, I.R. and S. Langton, 2006. How an animal's behavioural repertoire changes in response to a changing environment: a stochastic model. Behav. 143, 1563-1596.
Jensen, M.B. and R. Kyhn, 2000. Play behaviour in group-housed dairy calves, the effect of space allowance. Appl. Anim. Behav. Sci. 67, 35-46.
Jongman, E.C., I. Bidstrup and P.H. Hemsworth, 2005. Behavioural and physiological measures of welfare of pregnant mares fitted with a novel urine collection device. Appl. Anim. Behav. Sci. 93, 147-163.
Johnson, K.G., J. Tyrrell, J.B. Rowe and D.W. Petherick, 1998. Behavioural changes in stabled horses given non-therapeutic levels of virginiamycin, Equine Vet. J. 30, 139-142.
Johnson, S.R., E.G. Patterson-Kane and L. Niel, 2004. Foraging enrichment for laboratory rats. Anim. Welfare 13, 305-312.
Jørgensen G.H.M. and K.E. Bøe, 2007. A note on the effect of daily exercise and paddock size on the behaviour of domestic horses (*Equus caballus*). Appl. Anim. Behav. Sci. In Press.
Kiley-Worthington, M., 1984. Time-budgets and social interactions in horses: The effect of different environments. Appl. Anim. Behav. Sci. 13, 181-182.
Kimura, R., 1998. Mutual grooming and preferred associate relationships in a band of free-ranging horses, Appl. Anim. Behav. Sci. 59, 265-276.
Krohn, C.C., 1994. Behaviour of dairy cows kept in extensive (loose housing/pasture) or intensive (tie stall) environments. III. Grooming, exploration and abnormal behaviour. Appl. Anim. Behav. Sci. 42, 73-86.
Krohn, C.C. and M.D. Rasmussen, 1992. Dairy cows under extreme conditions; production, reproduction health and stayability. Beretning fra statens husdyrbrugsforsøg, Foulum, Denmark, p. 56.
Lammers, G.J. and G.P. Schouten, 1985. Effect of pen size on the development of agonistic behaviour in piglets, Netherl. J. Agric. Sci. 33, 305-307.
Le Neindre, P., 1993. Evaluating Housing Systems for Veal Calves. J. Anim. Sci. 71, 1345 1354.
Lidfors, L., 1989. The use of getting up and lying down movements in the evaluation of cattle environments, Vet. Res. Commun. 13, 307-324.

Ludes, E. and J.R. Anderson, 1996. Comparison of the behaviour of captive white-faced capuchin monkeys (*Cebus capucinus*) in the presence of four kinds of deep litter. Appl. Anim. Behav. Sci. 49, 293-303.

Ludes-Fraulob, E. and J.R. Anderson, 1999. Behaviour and preference among deep litters in captive capuchin monkeys (*Cebus capucinus*). Anim. Welfare 8, 127-134.

Mal M.E., T.H. Friend, D.C. Lay, S.G. Vogelsang and O.C. Jenkins, 1991. Behavioural Responses of mares to short-term confinement and isolation. Appl. Anim. Behav. Sci. 31:13-24.

Manninen, E., A.M. de Passille, J. Rushen, M. Norring and H. Saloniemi, 2002. Preferences of dairy cows kept in unheated buildings for different kind of cubicle flooring, Appl. Anim. Behav. Sci. 75, 281-292.

Marriner, L.M. and L.C. Drickamer, 1994. Factors influencing stereotyped behaviour of primates in a zoo, Zoo Biol. 13, 267-275.

Mason, G.J., 1991. Stereotypies: a critical review. Anim. Behav., 1991, 41, 1015-1037.

McAfee, L.M., D.S. Mills and J.J. Cooper, 2002. The use of mirrors for the control of stereotypic weaving behaviour in the stabled horse, Appl. Anim. Behav. Sci. 78, 159-173.

McDonnell, S.M., 2000. Reproductive behavior of stallions and mares: comparison of free-running and domestic in-hand breeding. Anim. Reprod. Sci. 60-61, 211-219.

McGreevy, P.D., P.J. Cripps, N.P. French, L.E. Green and C.J. Nicol, 1995a. Management factors associated with stereotypic and redirected behaviour in the thoroughbred horse, Equine Vet. J. 27, 86-91.

McGreevy, P.D., N.P. French and C.J. Nicol, 1995b. The prevalence of abnormal behaviours in dressage, eventing and endurance horses in relation to stabling. Vet. Record. 137, 36-37.

Meunier-Salaün, M.C., M.N. Vantrimponte, A. Raab, and R. Dantzer, 1987. Effect of floor area restriction upon performance, behaviour and physiology of growing-finishing pigs. J. Anim. Sci. 64, 1371–1377.

Mignon-Grasteau, S., A. Boissy, J. Bouix, J.M. Faure, A.D. Fisher, G.N. Hinch, P. Jensen, P. Le Neindre, P. Mormede, P. Prunet, M. Vandeputte and C. Beaumont, 2005. Genetics of adaptation and domestication in livestock, Livest. Prod. Sci. 93, 3-14.

Mills, D.S, 2005. Repetitive movement problems in the horse. In: D. Mills and S. McDonnell (Eds.) The domestic horse, the evolution, development and management of its behaviour. Cambridge University Press. 212-225.

Mills D.S. and K. Davenport, 2002. The effect of a neighbouring conspecific versus the use of a mirror for the control of stereotypic weaving behaviour in the stabled horse. Anim Sci 74, 95-101.

Mills, D.S, S. Eckley. and J.J. Cooper, 2000. Thoroughbred bedding preferences, associated behaviour differences and their implications for equine welfare. Anim. Sci. 70, 95-106.

Morgan, K.N. and C.T. Tromborg, 2007. Sources of stress in captivity. Appl. Anim. Behav. Sci. 102, 262-302.

Napolitano, F., G. de Rosa, F. Grasso, C. Pacelli and A. Bordi, 2004. Influence of space allowances on the welfare of weaned buffalo (*Bubalus bubalis*) calves. Livestock. Prod. Sci. 86, 117-124.

Nevison, C.M., J.L. Hurst and C.J. Barnard, 1999. Why do male ICR (CD-1) mice perform bar-related (stereotypic) behaviour?, Behav. Process. 47, 95-111.

Nicol, C., 1999. Understanding equine stereotypies. Equine Vet. J. Suppl. 28, 20-25.

Pearce, G.P. and A.M. Patterson, 1993. The effect of space restriction and provision of toys during rearing on the behaviour, productivity, and physiology of male pigs, Appl. Anim. Behav. Sci. 36, 11-28.

Pederson, G.R., E. Sondergaard, and J. Ladewig, 2004. The influence of bedding on the time horses spend recumbent, J. Equ. Vet. Sci. 24, 153-161.

Plowman, A.B., N.R. Jordan, N. Anderson, E. Condon and O. Fraser, 2005. Welfare implications of captive primate population management: behavioural and psycho-social effects of female-based contraception, oestrus and male removal in hamadryas baboons (*Papio hamadryas*). Appl. Anim. Behav. Sci. 90, 155-165.

Polley, C.R., J.V. Craig and A.L. Bhagwat, 1974. Crowding and agonistic behaviour: a curvilinear relationship?. Poult. Sci. 53, 1621-1623.

Price, E.O., 1984. Behavioural aspects of animal domestication. Q. Rev. Biol. 59, 1-32.

Raabymagle, P. and J. Ladewig, 2006. Lying behaviour in horses in relation to box size. J. Equ. Vet. Sci. 26, 11-17.

Regal P.J and M.S. Connolly, 1979. Social influences on biological rhythms. Behav. 72, 171-198.

Rodenburg, T.B. and P. Koene, 2003. Comparison of individual and social feather pecking tests in two lines of laying hens at ten different ages. App. Anim. Behav. Sci. 81, 188-148.

Sherwin, C.M, E. Haug, N. Terkelsen and M. Vadgama, 2004. Studies on the motivation for burrowing by laboratory mice, Appl. Anim. Behav. Sci. 88, 343-358.

Sneddon, I.A. and Beattie, V.E., 1995. Improving the welfare of pigs, Irish J. Psychol. 16, 418-425.

Sørensen, D.B., T. Krohn, H.N. Hansen, J.L. Ottesen and A.K. Hansen, 2005. An ethological approach to housing requirements of golden hamsters, Mongolian gerbils and fat sand rats in the laboratory - A review. App. Anim. Behav. Sci. 94, 4181-195.

Spinka, M., 2006. How important is natural behaviour in animal farming systems? Appl. Anim. Behav. Sci. 100, 117-128.

Stoinski, T.A., M.P. Hoff, K.E. Lukas and T.L. Maple, 2001. A preliminary behavioural comparison of two captive all-male gorilla groups, Zoo Biol. 20, 27-40.

Stolba, A. and D.G.M. Wood-Gush, 1989. The behaviour of pigs in a semi-natural environment. Anim. Prod. 48, 419-425.

Stolba, A., N. Baker, and D.G.M. Wood-Gush, 1983. The characterisation of stereotyped behaviour in stalled sows by informational redundancy. Behaviour, 87, 157 181.

Studnitz, M., K.H. Jensen, E. Jørgensen and K.K. Jensen, 2003a. The effect of nose-ringing on exploratory behaviour in gilts. Anim. Welfare 12, 109-118.

Studnitz, M., K.H. Jensen and E. Jørgensen, 2003b. The effect of nose rings on the exploratory behaviour of out door gilts exposed to different tests. Appl. Anim. Behav. Sci. 84, 41-57.

Thorne, J.B., D. Goodwin, M.J. Kennedy, H.P.B. Davidson and P. Harris, 2005. Foraging enrichment for stabled horses: practicality and effects on behaviour. Appl.Anim. Behav. Sci., 94, 149-164.

Thimonier, J., 1981. Control of seasonal reproduction in sheep and goats by light and hormones. J. Reprod. Fertil. Suppl. 30, 33-45.

Tuyttens, F., F. Wouters, L. Duchateau and B. Sonck, 2004. Sows prefer to lie on a prototype lying mattress rather than on concrete. In: L. Hänninen and A. Valros (Eds.) Proceedings of the 38th International Congress of the ISAE, Helsinki, Finland, 4-7 August.

Van Putten, G. and J. Dammers, 1976. A comparative study of the well-being of piglets reared conventionally and in cages. Appl. Anim. Behav. Sci., 2, 339-356.

Veasey, J.S., N.K. Waran, and R.J. Young, 1996. On comparing the behaviour of zoo housed animals with wild conspecifics as a welfare indicator, using the giraffe (*Giraffa camelopardalis*) as a model. Anim. Welfare. 5, 139-153.

Waring, G.H., 1983. Horse Behaviour: Behavioural Traits and Adaptations of Domestic and Wild Horses, Including Ponies. Noyes Publications, Park Ridge, NJ. 292 p.

Webster, J., C. Saville and D. Welchman, 1986. Improved Husbandry Systems for Veal Calves, Farm Animal Care Trust, London, UK.

Weinert, D., P. Fritzsche and R. Gattermann, 2001. Activity rhythms of wild and laboratory golden hamsters (*Mesocricetus auratus*) under entrained and free-running conditions, Chronobiol. Int. 18, 921-932.

Wiedenmayer, C., 1997. Causation of the ontogenetic development of stereotypic digging in gerbils. Anim. Behav. 53, 461-470.

Wiegand, R.M., H.W. Gonyou and S.E. Curtis, 1994. Pen shape and size: effects on pig behavior and performance. Appl. Anim. Behav. Sci. 39, 49-61.

Wiepkema, P.R., D.M. Broom, I.J.H. Duncan, and G. van Putten, 1983. Abnormal Behaviours in Farm Animals. Brussels: Report of the Commission of the European Communities.

Willard, J.G., J.C. Willard, S.A. Wolfram, and J.P. Baker, 1977. Effect of diet on cecal pH and feeding behaviour of horses. J. Anim. Sci. 45, 87-93.

Wood-Gush, D.G.M. and R.G. Beilharz, 1983. The enrichment of a bare environment for animals in confined conditions. Appl. Anim. Ethol. 10, 209-217.

Wood-Gush, D.G.M. and K. Vestergaard, 1989. Exploratory behaviour and the welfare of intensively kept animals. J. Agric. Ethics. 2, 161-169.

Young, R.J., J. Carruthers and A.B. Lawrence, 1994. The effect of a foraging device (The 'Edinburgh Foodball') on the behaviour of pigs. Appl. Behav. Anim. Sci. 39, 237-247.

Building and process technology requirements in horse husbandry systems: current investigations and developments furthering animal welfare and environmental protection

F.-J. Bockisch[1], P. Kreimeier[1], G. Hoffmann[1], T. Hohmann[1], W. Bohnet[2] and U. Brehme[3]
[1]Institute of Production Engineering and Building Research, Federal Agricultural Research Centre (FAL), Bundesallee 50, 38116 Braunschweig, Germany
[2]Centre for Animal Welfare - Institute of Animal Welfare and Behaviour, University of Veterinary Medicine, Bünteweg 2, 30559 Hannover, Germany
[3]Leibniz Institute for Agricultural Engineering Bornim (ATB), Max-Eyth-Allee 100, 14471 Potsdam, Germany

Abstract

This overview presentation addresses the design and performance of different horse husbandry systems and their effect on welfare and environmental parameters. The focus is on current and recent investigations carried out at our research facilities in cooperation with other institutes. The results are set in the context of the existing literature in this area. Our research facility comprises 6 single boxes for six horses (one box per horse) and a group housing system for 6 horses. Together, 12 Hanoverian standard-bred mares are accommodated and studied. Each treatment unit (single boxes and group) is directly connected with corresponding paddocks. For additional movement, there is a computer-operated free walking unit and pastures. Parameters monitored in several investigations are: quantitative behaviour using continuous video monitoring and analysis, online measurement of the heart rate, online registration of activity and resting by ALT-Pedometers, time and frequency of feed intake (roughage and concentrate feed) using a computerised feeding system, continuous monitoring of house climate and weather parameters, analyse of water and soil in the paddocks for emissions to the environment. The results show that group housing systems enable a reduction in the ratio of feeding place to animals to 1:3; increasing the intervals between concentrate feeding times results in a deterioration in heart frequency variability. Investigations of different combinations of space and movement possibilities show that horses require more then 45 m^2 per horse. 80% to 90% of the urinating activities of the horses take place on straw litter within the stable building rather than outside in the paddock area greatly reducing the risk of the contamination of soil and ground water.

Keywords: group husbandry, feeding technique, movement activity, environmental protection in connection with paddocks, heart beat frequency, ALT-pedometers

Initial situation

Against the background of ever-increasing numbers of horse and changing types of use, especially in Germany, consideration of species-appropriate, environment-friendly and advance process technique husbandry practices for horses is a topic of current importance. The number of horses and ponies kept in Germany in the past 35 years has increased to more than one million animals (Figure 1). World-wide, horse numbers have tripled over the same period according to World Conservation Monitoring (an international database for natural protection under the auspices of the United Nations). In Germany, horse husbandry presents a significant market for agriculture. Horses and ponies in Germany consume more than 1.6 million tons of feed grains and about 1.8 million tons of hay and straw each year. Horses are also a significant source of employment with three to four horses for every workplace, and about 2.6 billion € are spent by riders, other users and breeders each year (German Riding Union - FN, 2001). In total, equine husbandry and related activities are responsible for a turnover of approximately 5 to 6 billion € in the German economy.

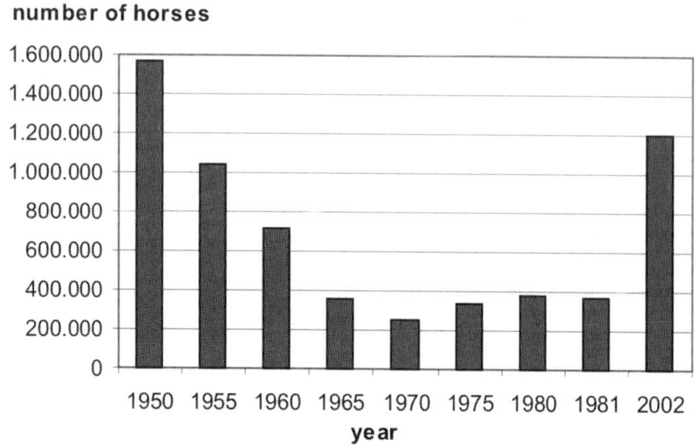

Figure 1. Development of the number of horses in Germany since 1950 (FN, 2001).

Regardless of how horses are used, a large percentage of the horses are kept in single boxes (typically 9 m^2) without free access to movement areas. Much of the accommodation is oriented to the needs of the people looking after the horses rather than to the horses themselves. Such facilities cannot fulfill the requirements appropriate to the husbandry of horses (Kreimeier and Bockisch, 1996). The consequences are locomotive and respiratory diseases as well as behaviour problems. Horses need several hours of free movement, fresh air, feed throughout the day and social contact to other horses. The free yard husbandry of groups furthers species appropriate husbandry.

Problem and goals

In addition to increasing emphasis on the protection of the environment, there is a significant trend toward species appropriate husbandry systems for welfare reasons. The free yard husbandry in groups is increasing in popularity primarily with recreational horseback riders. Single box husbandry with direct access to the yard is also an alternative for high performing horses or tournament horses used in sport. Studies conducted on trotters by Gerken *et al.* (1997) showed that group husbandry is suitable for high performance horses and provides better movement possibilities and less behavioural abnormalities (Table 1).

Other studies have focused on how well the needs of horses are met in group and individual free movement systems, and on the emissions to the environment associated with adjoining paddocks.

Table 1. Comparison of some criteria between group and single housing systems for trotting race horses.

Criteria	Housing systems	
	Group housing	Single housing
Individual feeding	more extensive	more easy
Possibility for movement	very good	insufficient
UV lighting	very good	insufficient
Creation of groups	take care of 'antagonism'	take care of 'antagonism'
Dev. from normal behaviour	<1% of observation time	3-5% of observation time
Performance	not influenced	not influenced

Research has, for example, focused on the effects of free yard husbandry with regard to the use of the functional areas. Studies of excretory processes provide information relevant to understanding the contamination of soil and leachate for further study questions.

For group husbandry, research has the focused on the influence of changes in the number of animal feeding places in relation to the use of the functional areas and behaviour within the horse groups. Different concentrate presentation methods (manual and automatic feeding) with and without waiting times lead to measurable stress differences in individually kept warm-blooded horses. A further study considered whether the movement needs of the horses can be met by open yard husbandry and which additional movement offerings could lead to an increase in movement activity and the even-temperedness of the horses.

In the various studies reported here, the following questions were addressed:

- which husbandry options and feeding situations are best suited for the health and well-being of the horses;
- are the natural needs (social contacts, movement and resting behaviour) met to an adequate extent for the horses;
- is there groundwater pollution from adjoining paddocks permanently used by horses.

The insights provided by this research project should contribute to an improvement of the open yard husbandry systems of horses.

Framework conditions for the different experiments

Experimental facilities

The horse testing facility of the Institute of Production Engineering and Building Research is located in the experimental station at the FAL Braunschweig. It was built as an unheated, non thermal insulated stable, fully constructed from wood with air conditioning carried out as a free ventilation system via wind protection nets, the roof and doors. The entire stable area has a surface from about 900 m². The stable is divided into two areas. In one part of the building there is a multi-room group open yard area for six horses, and in the other area, six boxes are located with adjoining sand paddocks, whereby one paddock and two boxes are commonly used by two horses (Figure 2 and 3).

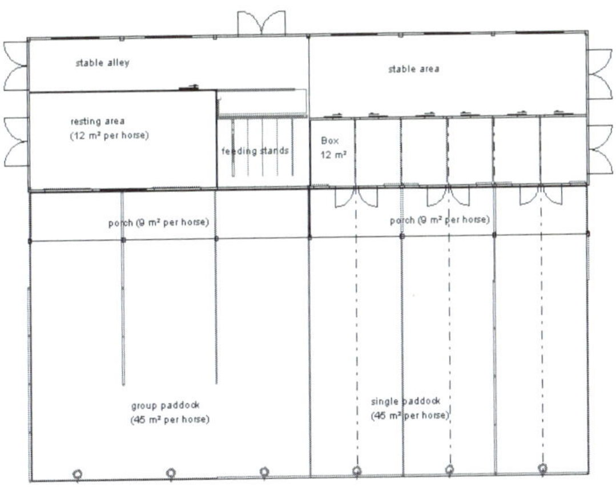

Figure 2. Ground plan of the research stable/research husbandry system.

Figure 3. Research stable for the investigations on horse husbandry systems (left site: group husbandry system with paddocks; right site: single husbandry system with paddocks).

Under group husbandry, the six horses have a covered straw litter (lying) area, a paved, roofed area, a sand paddock and six feeding stands. The straw litter area has two broad entry and exit pathways leading to the paddock. A trough and a salt stone are located in the lying areas, and three further troughs are located at the head of the paddock. The troughs can all be heated. The sand paddock is divided by two barrier elements. The horses must go around these barriers to reach the feeding areas. In the feeding areas, feeding takes place with an automatic hay slider and an automatic electronic identification concentrate feeding system.

In front of the boxes where the horses are kept in pairs, there is also a roofed and paved area with an adjoining sand paddock permanently used by horses in pairs. A salt stone and a drinking trough are located in each box. Furthermore, a drinking trough is located at the end of the three paddocks. For some studies, the yards were divided into single paddocks for a certain period of time by the creation of a mobile electrical gate. Through the division of the yard, each drinking trough is used by two horses. An automatic feeding device for concentrate feed is located in the box with straw litter and the automatic hay slider is integrated directly into each box door.

Feeding takes place in both group and individual husbandry via an automatically controlled hay slider and concentrate dispenser. There are six feeding times per day in both the group and individual husbandry treatments. 500 g oats are fed per horse with a concentrate dispenser three times per day (7 a.m., 1 p.m. and 6:30 p.m.). In addition, the horses receive hay via a time-controlled automatic feeder with a total feeding time of 270 minutes which is divided into three mealtimes of 90 minutes each (Feeding at 5:10 a.m., 4:40 p.m. and 8:50 p.m.). The horses also receive 150 g mineral feed per horse.

The group husbandry facility enables individual animal concentrate and roughage feeding. The animals are equipped with collars with transponders for electronic animal recognition. Each horse is assigned a feeding time per day and a defined concentrate feed amount, which is distributed by the computer over 20 hours a day (four hours are maintained as reserve time). The horse is recognised via its transponder when it enters the feeding area and the amount of feed allowed is checked. If the animal is due feed, then the hay slider slides up and opens the feeding gate or lets oats out of the dispenser. Depending on the pre-settings, the horses can freely choose their feeding areas or are assigned to a certain feeding stand. The computer can constantly monitor whether the horses fully use their feeding times or not. All stable tasks are completed in the morning, whereby the single boxes and resting areas of the group husbandry are cleaned daily and new litter is distributed. A protocol of the manure removal for each of the stable areas is kept daily. For additional exercise possibilities the horses have a pasture, a non-grass yard or a horse walker is available for daily movement.

Material and methods

Research is based on 12 test horses (which always are present at the same time), which are two- to three-year old German warm-blooded horses kept in the facility for six months; three months in group husbandry and three months in single boxes with paddocks. The horses are all female, all are of the Hanover breed and are all provided to the facility by their breeders. Following the study they are returned to their owners. Since the horses are young, it is easier to acclimatise them to the stable conditions and own previous experience has shown that younger horses react more openly to experimental treatments (these findings are not published before).

A permanent video monitoring system continually follows the activities of the horses so that their behaviour can be analysed and evaluated (Figure 4). In the stable segment of the group and individual husbandry overall 8 color video cameras (in each husbandry system 4) are installed in each section to monitor the entire stable and external area. One multiplexer bundles the pictures from four cameras. To support the night pictures, emergency lighting is on during the dark. The video are subsequently digitalised and saved on an external hard disc drive. The viewing and evaluation of the digitalised video data on the PC is conducted with the program 'interact' (brand: Mangold Co.), a specialised program for behaviour analysis.

To monitor treatment effects on stress, the heart frequency variability of the horse is measured. The measurements take place via two electrodes fastened via a belt around the horse's midsection together with a transmitter and a clock which initially stores the measurement data (Figure 5). Subsequently, the data are transferred via an infrared interface to a computer, where they are evaluated and processed with appropriate analysis software. In addition to the software from the Polar company,

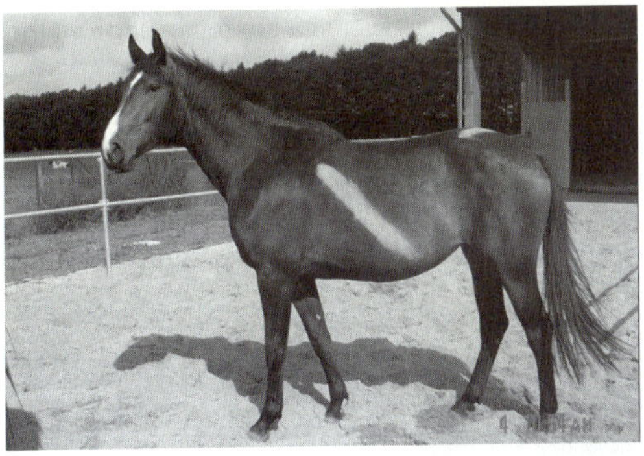

Figure 4. Example of a signed horse that secures the identification for the video analysis.

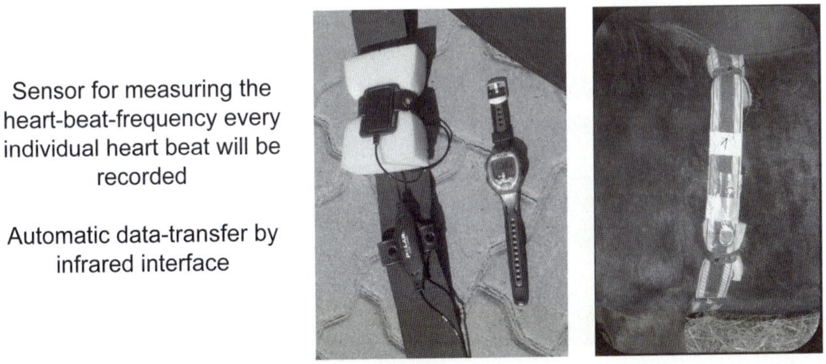

Figure 5. Measuring of the variability of the heart beat frequency.

a special analysis program for heart frequency variability from the Biomedical Signal Analysis Group (Department of Applied Physics; University of Kuopio, Finland) is used. To analyse the heart frequency variability, different time periods and frequency range parameters are used correlating to a level of the parasympathic nerve stimulation. Reduced stimulation, such as for example through stress, leads to a drop in the measured values, so that the psychological stress can be quantified. With the help of power spectral analysis increased stress can be documented.

As a supplementary measurement parameter, faeces samples can be taken from the horses and the concentration of cortisol metabolites can show how highly stressed the animals are. An enzyme immunoassay was developed to measure the excreted products. This measures a group of metabolites from cortisol which is created in the metabolisation processes in the body of the animal. These faeces analyses are carried out in cooperation with the FAL (Federal Agricultural Research Centre) Institute of Crop and Grassland Science in Braunschweig and the Institute of Biochemistry, University of Veterinary Medicine, Vienna.

The movement activity and laying times of the horses are recorded with pedometers (a type of step counter) which are attached 24-hours a day to a horse's hind leg. From the pedometer data, it can be seen how much time each horse spends moving around and lying down, with an automatic distinction between lying on the stomach and the side. The pedometers are called ALT-(**A**ctivity, **L**ying, **T**emperature) pedometers. Movement impulses are measured on the hind leg of a horse, where per second a maximum of two impulses are registered (Figure 6). Whether, and in which position, a horse lies, is monitored every 15 seconds with a location sensor. Based on these pre-settings, a calculation of the movement activity and total lying time in minutes is possible for every day. The ambient temperature at the horse's leg is measured every 15 minutes, but is not regarded further in this study. The ALT-pedometers are automatically consulted in a four-hour cycle and the stored data are transferred to a central computer. The data are removed from the storage and placed into the spread sheet program Excel for further processing.

Since the weather also has an influence on animal behaviour as well as on water and soil contamination, weather data are monitored over the entire course of the experiment in a near-by weather station. This includes a microprocessor weather station to measure and document the temperature, relative humidity, dew point, barometric pressure, solar energy, sunshine duration, precipitation, wind direction and wind speed.

ALT-Pedometer (Activity and Lying periods, Temperature at the leg)

These sensors are recording the time periods in which the horse is in movement and in lying position

Figure 6. Recording of locomotion activity and lying behaviour of horses with ALT-pedometers.

Examples for results

Influence of a changed animal to feeding place ratio in a group husbandry system

Here the influence various animal to feeding place ratios (1:1, 2:1 and 3:1) on the behaviour of the horse and its health was studied. Depending on the feeding technology, crude forage was given up to 20 times per day to emulate the continuous intake of food at pasture. Overall, the health of the 48 Hanovarian mares (over the entire investigation period) during their at least six month stays was very good. Over the two years of the study, there was not a single case of colic, and no horse had to leave the experiment prematurely. The nutritional status was good in all horses throughout the study periods, and the expected weight gain in the case of pregnant mares was also achieved.

In the evaluation of the feeding data it was striking that the individual horses showed significant preferences for a particular feeding place (Figure 7). In the change of animal-feeding place ratio from 1:1 to 2:1 it was already evident that no specific place was chosen (Figure 8) but apparently the next free slot was used. The video documentation shows that the behaviour changed in the case of increasing animal to feeding place ratio in that the mares apparently spent more time standing around the feeding places and that more defecation took place there (Kreimeier *et al.*, 2005).

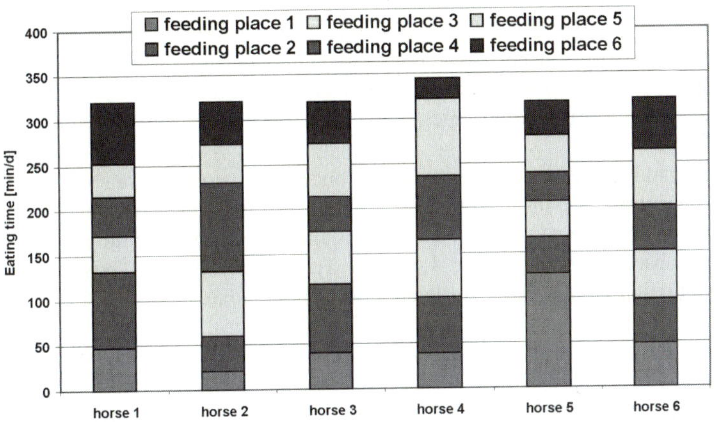

Figure 7. Daily eating time per horse and in different feeding places in group husbandry at a ratio of horse to feeding places of 1:1.

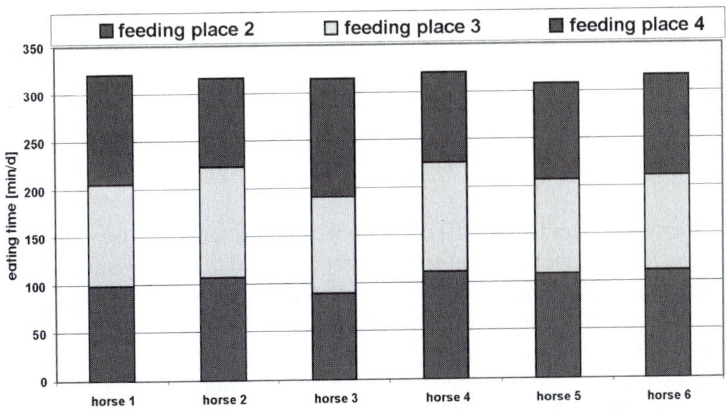

Figure 8. Daily eating time per horse and in different feeding places in group husbandry at a ratio of horse to feeding places of 2:1.

Effects of different concentrate feeding techniques to the horses

A further study considered the stress on horses during the feeding of concentrates. In the horse husbandry with single boxes and manual feed presentation, particularly with concentrates, the problem often emerges that the horses become agitated because all the feed is not offered at the same time causing risk of injury and stress for the animals. In this study, a behaviour analysis was linked with an analysis of the heart frequency variability in order to quantify a possible stress situation during the offering of concentrates with various feed application techniques and time shifts. In the case of the concentrate offering, an automatic, stationary concentrate dispenser (brand: Weinsberger Co.) was used. Each feeding station is equipped with a dispenser. The concentrate offering was time-controlled and occurred at all feeding stations simultaneously.

The behavioural expressions shown by the horses were monitored with video cameras and placed in relation to the values of the analysis of heart frequency variability. The behaviour was divided into basic and additional activities. The basic activities are feeding, drinking, standing and lying. The additional activities are typical behavioural activities that the horses show when they receive concentrates manually, such as, for example, pawing, pacing, etc.

The study design consisted of three treatments:
Treatment I: 10 x daily concentrate feeding with automatic simultaneous feeding technique.
Treatment II: 3 x daily concentrate feeding with automatic simultaneous feeding technique.
Treatment III: 3 x daily concentrate feeding, manual, time-shift from horse to horse from 30 seconds, that means waiting times from 0, 0.5, 1.0, 1.5, 2.0 and 2.5 minutes.

In treatments I and II, the horses received their concentrates simultaneously with a time-controlled automatic concentrate dispenser. In treatment III, the manual concentrate presentation was carried out with delays. This means that the first horse in the experimental stable received its concentrate immediately, subsequent horses had to wait 0.5, 1, 1.5, 2 or 2.5 minutes. Thus treatment III simulated the feeding situation in most horse husbandry with single boxes.

In treatment I, the heart frequency mean was lower and the value of the heart frequency variability analysis (the High Frequency Power – HF-Power) was higher than for treatment II, which represents a reduction in stress. In treatment III, the average heart frequency increased with increased waiting time while in contrast the HF-Power reduced up to a waiting time of one minute and then increases slightly. In the behavioural analysis, behavioural responses are scarcely detectable for treatments I and II. In treatment III, the number of expressions of behaviour responses increases with increased waiting time. The following behavioural responses were shown the most: pausing, pawing, threatening, head-butting, pacing. Both the higher values in heart frequency measurement and the behavioural analysis, as well as the lower values in high frequency power (Figure 9 and 10), indicates an increased stress level for the animals due to a waiting time while concentrates are offered manually. With automatic concentrate feeding technique (Variations/treatments I and II) the ten times a day concentrate feeding regime reduced stress.

In summary it can be said, that the offering of concentrates ten times per day using a concentrate dispenser has the lowest psychological stress effect in regard to the stress for horses in this study. It must be particularly emphasised that a stress level observed through the heart frequency variability analysis is not always in accordance with behavioural activity such as pawing or threatening. In practice this means that the automated concentrate offering technique is only less stressful for horses than manual feeding if all horses receive their feed at the same time. A repeated, automatic and simultaneous presentation of concentrates is thus an improvement in the husbandry conditions with regard to the health of horses held in single box systems (Hohmann *et al.*, 2005, 2006; Bockisch *et al.*, 2007).

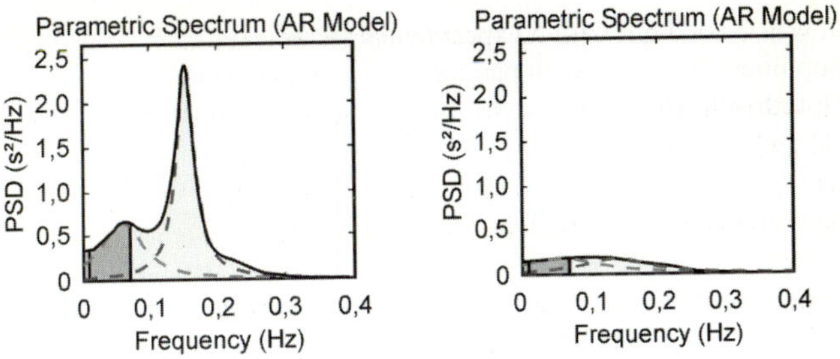

Figure 9. Example from one horse for the PSD (Power Spectral Density) in dependence of the heart beat frequency by concentrate feeding without (left site) and with (right site) waiting time; with waiting time there are lower values in HF-Power and this means normally more stress.

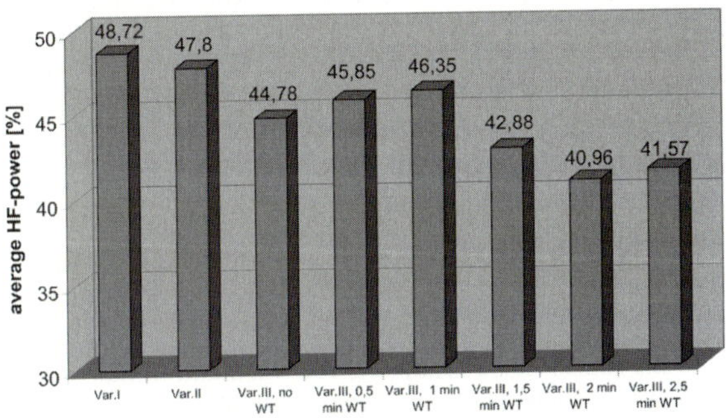

Figure 10. Average HF-Power of six horses by different kinds (variations/treatments) of concentrate feeding.

Influence of different movement and space offerings on the movement behaviour of horses

The effects of different movement offerings on the stress behaviour and the movement activity of horses in free yard husbandry were also studied in one project. The movement behaviour off 24 horses was studied in single and group husbandry systems with paddocks, while the group husbandry was augmented with three additional movement offerings (pasture, soil yard, horse walker). Here, the movement needs of horses in single and group husbandry systems, each with paddocks can be examined. Research looks at the effects of the form or intensity of movement on the well-being of the animal. During the study period, ALT-pedometers were attached to the hind legs of the horses to monitor their movement activity; additionally the behaviour was recorded with video technique. The stress on the horses was documented through measurements of the heart frequency variability and the determination of the cortisol metabolites in faeces.

Frentzen (1994) was able to show in her studies at the Institute of Farm Buildings of the FAL Braunschweig, that the feeding frequency and the distance to the feeding slots had a greater influence on the distance horses traveled within the husbandry system than the possibility for free movement within the husbandry system. The research reported here looked at:

- if horses held only in a group free yard husbandry (with paddocks) have enough incentives for movement; and
- which additional movement offerings could lead to an increase in movement activity.
-

The investigation was carried out with 5 experimental treatments:
Treatment I: Group housing system with paddock (GH) with pasture.
Treatment II: GH with automatic exercise facility (horse walker).
Treatment III: GH with soil run.
Treatment IV: GH.
Treatment V: Single housing system with paddocks (SH).

The horses spent between 100 and 150 minutes per day moving around during the entire study period (Hoffmann *et al.*, 2006). The movement activity of the horses during the individual study is illustrated in (Figure 11). The least movement activity was shown in the single free run husbandry (Treatment V), with an average of 102 minutes of movement per day. The group husbandry with different functional areas and paddocks with and without additional movement offerings led to an increase in the average movement time to more than 120 minutes per day. In treatment II, with one hour of movement in a horse walker facility, measurement data was corrected since 'forced' movement led to a strong increase in recorded activity. The number of movement impulses in this hour was replaced with a mean of the other 23 hours of each day. The evaluation of the results following the correction yielded an average movement time of 122 minutes per day in treatment II. The results in the group housing treatment without additional movement opportunities (Treatment IV) were similar. The horses did not leave the stable complex in this treatment and moved for an average of 125 minutes per day. The additional movement on a soil run (Treatment III) also led to only a small increase of the movement time of 132 minutes per day. An additional movement opportunity in the form of a two hour pasture visit (Treatment I) resulted in an average movement activity of 150 minutes per day. In the fourth horse group, a three week additional variation occurred in which the horses from the group husbandry had a paddock available which was two times as large as the common paddock area at the stable. For organisational reasons this variation was not applied to the other horse groups. The duration of movement monitored with the ALT-pedometer was 114 minutes on average. This time is only one minute more than the movement time of this horse group in the variation of the usual group husbandry (GH – treatment IV) with paddocks, without additional movement opportunity.

The determination of the heart frequency variability and the cortisol metabolites in faeces showed that stress indicators were lowest in the horses in the treatment group which had two hour stays at pasture and had one hour movement in the horse walker. The measurement methods showed a stress increase over these in the group husbandry with two hours free run on a soil area as well as in the treatments comprising single and group housing without additional movement opportunities.

In general, one can conclude from this study that the horses in the group husbandry with paddocks can move more than in the usual single box husbandry with paddocks. But with a maximum of

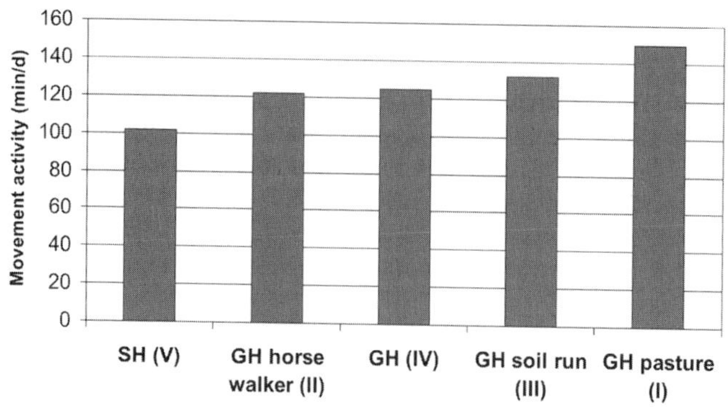

Figure 11. Movement activity in different kinds (treatments) of housing systems with paddocks.

2.5 hours of movement per day, movement is much less than under free-range conditions. Under near-to-nature conditions, the horses moved in a social unit for 12 to 16 hours per day. Under such circumstances, the feed consumption is characterised by a purposeful, slow progression for about 60% of the daily activity interspersed by standing. (Pirkelmann, 2002). In the observations reported here of horse behaviour, it can be concluded that the animals tend towards inactivity. Even in the relatively large sand paddocks a great deal of time is devoted to dozing. Voluntary movement is only rarely observed and the animals always chose the shortest distance from the resting areas to the feeding areas, whereby through the group, a mutual urge to play emerged and thus more movement was evident.

In order to increase the movement activity, an incentive is needed for the horse to move. The group husbandry of horses with different functional areas and paddocks offers a certain motivation for movement but is not comparable with the movement behaviour of horses under free-range conditions.

The results of the study show that additional movement opportunities in the form of a soil run, horse walker and pasture increase the daily movement period. Enlargement of the sand paddocks adjoining the stable in group husbandry with different functional areas had no effect on the movement behaviour of the horses.

Investigation on year-round used paddocks in horse husbandry systems with a focus on the use of the functional areas and environmental aspects

The studies sought to provide information on how the horses use each functional area. Standing was much more evident than movement or lying. Group husbandry most effectively fulfills the movement and social needs of the horses. Horses spent more time in roofed areas, boxes or feeding areas or in lying areas at the expense of time in yards. With regard to defecation, horses tended to defecate mainly in the boxes and in the front part of the paddocks. They urinate mainly in the littered boxes or in the lying area (Figure 12). This shows that in these areas simple construction measures can be used to reduce the risk of emissions to water and while increasing the area allocated to open yards.

The goal of the environmental studies is to establish the level of pollution emissions to soil and water where highly permeable constructed floors are used (Figure 13). The ammonium, nitrate, phosphate and potassium concentrations in leachate were analysed as well as biological parameters (total bacteria count, coliform bacteria and *E. coli* bacteria).

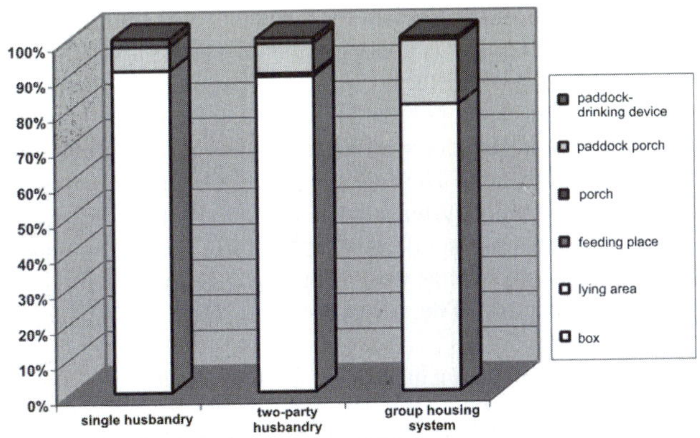

Figure 12. Relatively distribution of the urinating activities in the different function areas of husbandry systems with paddocks.

Figure 13. Facility (like a lysimeter system) for investigation to the environmental protection (water and soil) in paddocks.

Samples were taken using collection containers located in the free run areas of the feeding places under group husbandry. In this usage area frequent defecation can be observed. Results show that water from most of these collection sites had a annual mean nitrate concentration below 50 mg/l, i.e. the general guideline limit for ground water.

Conclusions

In addition with other recorded parameters and results from our own investigations and in consideration to results of the literature, in the first passage of this conclusion, there are named some complex aspects and implications.

Compared to stable systems it is clear that the horses in group husbandry with paddocks are very calm and even-tempered. They can fulfill most basic needs such as social contact, mutual hide care, freedom of movement and can pursue a relatively natural herd behaviour. Securing of need-appropriate care and the stress free integration of animals lower in the hierarchy presents challenges for the division of space, the equipping of the stable, and for stable personnel. With regard to freedom and motivation of movement, group husbandry with free yard access is the most appropriate form of husbandry after pasture-keeping. In each husbandry system the horse, in contrast to pasture keeping, must be motivated to move. In single free yard husbandry systems, the horses can feed and rest undisturbed but they have no partners for social contacts, which help as additional movement incentives. Pair husbandry with free run is the most appropriate for the needs of the horse in regard to social and resting behaviour. The nutritional intake is good, but the number of times the automatic feeders is used is very high. Group husbandry offers the horses a good possibility for free movement and social contacts, but does not satisfy their resting needs. The nutritional uptake is good, but varies strongly between the horses. Each of the husbandry systems tested here is, with small changes, are suitable for species appropriate horse husbandry. They are a significant improvement to conventional single box husbandry. In all of the husbandry systems the care, the feed, the water, stable hygiene and the manner of dealing with the horses plays a decisive role. Horse husbandry must be adapted to the natural needs of the horse. The physical as well as the psychological health of the horse must be given the highest priority irrespective of the use of the horse. Here feeding plays a large role.

In the following, there are the special conclusions for the four experiments, which are described in this paper:

1. In group husbandry a reduction of the animal-feeding place ratio has a negative effect on the movement and resting behaviour of the horse. Horses higher in the hierarchy often block the feeding places after the meal or push other horses out of the feeding area. Although the nutritional

requirements of the horse are not endangered, it would be better to provide each horse with a feeding place in order to eliminate stress.

2. With regard to the concentrate feeding technology used, a delay in feeding apparently has a psychologically negative effect on the horses. In the study, with increased waiting time, the number of negative behavioural expressions as well as the measurable stress level in the nervous system increased.
3. Studies of different movement offerings have shown that additional movement in the form of pasture access and horse walkers has a positive effect on the even-temperedness of the horses and an increase on movement activity. However, throughout the research, a low maximum movement period of three hours was observed. Horses held in the wild or under largely natural conditions, the search for food is a permanent incentive for movement. In contrast, a horse in a husbandry system receives adequate daily food without having to move for it. Therefore it requires additional incentives to move other than just the provision of a large paddock. Since horses are voluntarily willing to move to a limited extent, the goal is to increase their movement activity, since numerous studies show that this movement is very positive for horse health. A lack of movement can lead to significant problems in the circulatory system, to a lack of elasticity in the sinews, ligaments and joints and to an increased danger of ruptures. It is also responsible for inadequate self-cleaning of the lungs, for metabolism and hip mechanism problems as a consequence of inadequate blood circulation, for psychological problems, for displacement activity and movement stereotypes (Piotrowski and Kreimeier, 1998). The husbandry of horses in groups with paddocks and several functional areas leads to an increase in movement activities, but it is possible to increase willingness for movement activity further through structural changes. Developing this further is worthy of further research.
4. The initial results of the investigations of emissions to the environment reported here indicate that the risks are lower than thought at the outset. These risks could be reduced and managed through relatively simple alterations to the construction of some of the more frequented areas to reduce leaching contamination in the yard allowing increases in the size of the remaining yard areas.

References

Bockisch, F.-J., T. Hohmann, P. Kreimeier and W. Bohnet, 2007. Auswirkungen und Vergleich von automatischen und manuellen Kraftfuttervorlgetechniken in Einzelboxenhaltung von Warmblutpferden. '15. Arbeitswissenschaftliches Seminar VDI-MEG-Arbeitskreis Arbeitswissenschaften im Landbau' 5. und 6. März 2007 in Wien/Österreich. Landtechnische Schriftenreihe S. 13-19.

Deutsche Reiterliche Vereinigung (FN), 2001. IPSOS: Marktanalyse der FN zum Pferdesport. Pferdesportler in Deutschland., Warendorf. URL: http://www.wpsv.de/ipsos.htm (15.06.2006).

Frentzen, F., 1994. Bewegungsaktivitäten und -verhalten von Pferden in Abhängigkeit von Aufstallungsform und Fütterungsrhythmus unter besonderer Berücksichtigung unterschiedlich gestalteter Auslaufsysteme, Hannover, Diss.

Gerken, M., M. Kiene, P. Kreimeier and F.-J. Bockisch, 1997. Verhalten von Trabrennpferden in Gruppenauslaufhaltung und in Einzelhaltung, Aktuelle Arbeiten zur artgemäßen Tierhaltung 1996: Vorträge anläßlich der 28. Internationalen Arbeitstagung Angewandte Ethologie bei Nutztieren der Deutschen Veterinärmedizinischen Gesellschaft e.V. Fachgruppe Verhaltensforschung vom 7. bis 9. November 1996 in Freiburg/Breisgau, KTBL-Schrift 376. Auflage, Seite 132-143, KTBL-Schriften-Vertrieb im Landwirtschaftsverl.; Münster-Hiltrup.

Hoffmann, G., F.-J. Bockisch, P. Kreimeier and U. Brehme, 2006. Einfluss verschiedener Bewegungs- und Platzangebote auf das Bewegungsverhalten bei Pferden. KTBL-Schrift 448:157-164.

Hohmann, T., P. Kreimeier, F.-J. Bockisch and W. Bohnet, 2005. Auswirkungen unterschiedlicher Kraftfuttervorlagetechniken und -frequenzen auf die Herzfrepuenzvariabilität und das Verhalten von Warmblutpferden, Hoch, C., 7. Tagung Bau, Technik und Umwelt in der landwirtschaftlichen Nutztierhaltung 2005: 1.-3. März 2005 in Braunschweig. Seite 55-60, KTBL-Schriften-Vertrieb im Landwirtschaftsverl., Münster-Hiltrup.

Hohmann, T., P. Kreimeier, F.-J. Bockisch and W. Bohnet, 2006: Auswirkungen unterschiedlicher Kraftfuttervorlagetechniken und -frequenzen auf die Herzfrequenzvariabilität und das Verhalten von Warmblutpferden. KTBL-Schrift 448:119-125.
Kreimeier, P. and F.-J. Bockisch, 1996. Tierangepaßte Freizeitpferdehaltung, Landtechnik 51 (5): 284-285.
Kreimeier, P., S. Wrieske, F.-J. Bockisch and W. Bohnet, 2005. Einfluss der Änderung des Tier-Fressplatzverhältnisses bei Gruppenhaltung von Hannoveranerstuten auf die Nutzung der Funktionsbereiche und einige Verhaltensparameter, Hoch, C., 7. Tagung Bau, Technik und Umwelt in der landwirtschaftlichen Nutztierhaltung 2005: 1.-3. März 2005 in Braunschweig. Seite 415-420, KTBL-Schriften-Vertrieb im Landwirtschaftsverl., Münster-Hiltrup.
Piotrowski, J. and P. Kreimeier, 1998. Pferde-Auslaufhaltung, Bauen für die Landwirtschaft 1: 8-12.
Pirkelmann, H., 2002. Neuere Entwicklungen für pferdegerechte Haltungssysteme, Bayerische Landesanstalt für Tierzucht, Grub/München (Band 863): 2-14.